Palgrave Studies in Medieval and Early Modern Medicine

Series Editors
Jonathan Barry
Department of History
University of Exeter
Exeter, UK
Fabrizio Bigotti
Institute for the History of Medicine
Julius Maximilian University
Würzburg, Germany

CENTRE FOR THE STUDY OF
MEDICINE AND THE BODY
IN THE RENAISSANCE

The series focuses on the intellectual tradition of western medicine as related to the philosophies, institutions, practices, and technologies that developed throughout the medieval and early modern period (500-1800). Partnered with the Centre for the Study of Medicine and the Body in the Renaissance (CSMBR), it seeks to explore the range of interactions between various conceptualisations of the body, including their import for the arts (e.g. literature, painting, music, dance, and architecture) and the way different medical traditions overlapped and borrowed from each other. The series particularly welcomes contributions from young authors. The editors will consider proposals for single monographs, as well as edited collections and translations/editions of texts, either at a standard length (70-120,000 words) or as Palgrave Pivots (up to 50,000 words).

Vivian Nutton

Andreas Vesalius and his *Fabrica*, 1537–1564

Changing the World of Anatomy

palgrave
macmillan

Vivian Nutton
University College London
London, UK

ISSN 2524-7387 ISSN 2524-7395 (electronic)
Palgrave Studies in Medieval and Early Modern Medicine
ISBN 978-3-031-69564-3 ISBN 978-3-031-69565-0 (eBook)
https://doi.org/10.1007/978-3-031-69565-0

This Palgrave Macmillan imprint is published by the registered company Springer Nature Switzerland AG
The registered company address is: Gewerbestrasse 11, 6330 Cham, Switzerland

Acknowledgements

The origins of this book go back to 2010, when I received an email from Gerry Vogringic inviting me to assist him in deciphering and publishing what he believed to be Vesalius' own copy of the 1555 *Fabrica*. This, at first sight implausible, request has led me on a journey of discovery that has taken me across the globe, for not only was the volume heavily annotated by its author but it encouraged me to look more closely than ever before at some familiar as well as many unfamiliar aspects of the greatest of all anatomical books. This study both begins and ends with Gerry Vogringic and his discovery, and it is only right that he should come first in my expressions of gratitude. Likewise, I am grateful to Stuart Rose for access to his annotated copy of the Vesalian *Institutiones*, as well as to Philip Oldfield, Anne Dondertman and the staff of the Thomas Fisher Rare Books Library of the University of Toronto for allowing me to see this when it was on exhibition there in 2014 and to spend further time with the 1555 edition once it had been deposited there on loan.

My involvement with Vesalius quickly brought me into contact with many Belgians, whose friendship and advice have been crucial—in particular, Maurits Biesbrouck, Theo Dirix, a fellow taphophile, Theodore Godderis, Pascale Pollier, Omer Steeno, Francis Van Glabbeek, Rob Van Hee and Ann Van der Velde. The triennial meetings organised by this group in Zakynthos and Antwerp have allowed me to tap into their detailed knowledge of Vesalius as well as to meet a variety of scholars interested in him and his work. The Belgian Foreign Ministry also facilitated visits in 2014 to lecture to audiences in Copenhagen and Astana. The recent acquisition of the 1555 volume by the University of Louvain is a fitting indication of the Belgian regard for one of their greatest sons: Vesalius has come home.

All Vesalians must at some point make their way to Basle, where Martin Steinmann assisted my work in the University Library and Dominique Brancher invited me to deliver a lecture on Vesalius. I was also delighted to work with Dan Garrison and Malcolm Hast on their sumptuous English version of the 1543 *Fabrica* (and much else) that was published by the firm of Karger. Its head, Thomas Karger, and his successor, Gabrielle Karger, were generous hosts, and I learned much about printing from discussions with their staff, especially Herbert Weckerlin and Franziska Suter.

I am grateful to the University of Padua and Maurizio Rippa Bonati for an invitation to talk about Vesalius in rooms of the Bo' which he would have known. Fabiola Zurlini and the Studio Firmano also invited me to several conferences in Fermo, whose Biblioteca civica houses a hand-painted copy of the 1543 edition as well as one of the 1555 edition that belonged to Queen Christina of Sweden. Both the town and its library deserve to be much better known.

Closer to home I am grateful to comments and suggestions to audiences in Edinburgh, Glasgow, London, Oxford and Warwick and to the many friends who answered my questions, sent me books and articles or pointed me in unexpected directions, including Tawrin Baker, Lena Berger, Ron Blumenfeld, Piet Bols, Hélène Cazes, Lore Chumbley, the late Iain Donaldson, Dolores Iorizzo, Cindy Klestinec, David Lines, Ian Maclean, Hannah Marcus, Dan Margócsy, Domenico Meli. Michael Stolberg, Tilmann Walter, Andrew Wear, David Williams and audiences across Europe. Sachiko Kusukawa kindly showed me a proof of her own book on the *Fabrica* and discussed with me our complementary approaches to what happened in 1543 and after. Roger Gaskell read my sections on the printing of the 1543 and 1555 editions and corrected many errors of this typographical neophyte. Martin Kemp performed a similar service for my art-historical sections as well as allowing me to make use of his unpublished suggestion for the major illustrator of the *Fabrica*. Fabrizio Bigotti, Monique Kornell, Marjorie O'Boyle and David Soulier deserve special thanks for reading the whole manuscript and commenting in detail throughout. Their advice and expertise in very varied aspects of Vesaliana have improved the book enormously. Any errors or misunderstandings that remain are entirely my own.

Rhiannon Knol, the compiler of the 2024 Christie's sale catalogue of the 1555 edition, kindly arranged for me to use the images prepared for that sale. Lucy Kidwell, my editor, has provided timely advice and criticism, while my daughter, Alice Tyrell, helped resolve computer problems. My wife, as always, has provided constant encouragement and a sharp eye for errors and infelicities.

This book appears in a series sponsored by the Pisa Centre for the Study of Medicine and the Body in the Renaissance (CSMBR), of which I have had the honour to be the president. Its Director, Fabrizio Bigotti, and its administrator, Tomaso Pedrotti Dell'Acqua, have succeeded in a short time in creating an organisation that has gained world-wide recognition for its lectures and its *manifestazioni* as well as for its summer school that has brought together from around the world young scholars and senior experts in a wide range of disciplines. Their enthusiasm, as well as the very varied topics that have been discussed there, confirms that there is still a place for intellectual history in an increasingly scientific world. I trust that this volume lives up to the Centre's high standards.

The last fifteen years have seen many discoveries about Vesalius and his great book. This small offering has two aims: to bring new aspects of the man and his expertise in many different fields to wider attention and to encourage others to look beyond what was published in 1543. That would be a fitting tribute to a man who was never satisfied with what he had written and was constantly looking for something new.

Contents

A Brief Chronology of Early Sixteenth-Century Anatomy Writings

1502	Alessandro Benedetti, *Historia corporis sive Anatomice*
1520	Alessandro Achillini, *Anotomicae adnotationes*
1521	Jacopo Berengario da Carpi, *Commentaria in Anothomiam Mundini*
1522	Jacopo Berengario da Carpi, *Isagoge breves*
1524	Birth of Vesalius
1528	Galen, *Libri anatomici*
1536	Joachim Dryander, *Anatomia humani capitis*
	Johann Guinter, *Institutiones anatomicae*
	Andres Laguna, *Anatomica methodus*
	Niccolò Massa, *Anatomiae liber introductorius*
	Theophilus Protospatharius, *De corporis humani fabrica*
1537	Vesalius, *Paraphrasis,* ed. 1
1537	Vesalius appointed professor at Padua
1537/8?	Vesalius, *Paraphrasis,* ed. 2
1538	Vesalius, *Tabulae sex anatomicae*
1538	Johann Guinter and Andreas Vesalius, *Institutiones anatomicae*
1539	Johann Guinter, *Institutiones anatomicae,* ed. *2*
	Vesalius, *Epistola, docens venam axillarem dextri cubiti in dolore laterali secandam*
1541	Giambattista Canani, *Musculorum humani corporis picturata dissectio*
1542	Bassiano Landi, *De humana historia*
1543	Vesalius, *De humani corporis fabrica*
	Vesalius, *Suorum de humani corporis fabrica librorum epitome*
	Vesalius, *Von der menschen Cörpers Anatomey*
1544	Vesalius enters imperial service

1544	John Caius, *Libri aliquot graeci*
1545	Charles Estienne, *De dissectione partium corporis humani*
	Thomas Geminus, *Compendiosa totius anatomiae delineatio*
	Vesalius, *Epistola, rationem modumque propinandi radicis Chynae decocti....pertractans*
1547	Christopher Langton, *A very brefe treatise*
1551	Vesalius, *Anatomia deudsch; ein kurzerAuszug der Beschreibung aller Glider menschlichs Leybs.*
1552	Vesalius, *De humani corporis fabrica*, Lyon edition
1555	Jacobus Sylvius, *In Hippocratis et Galeni physiologiae partem isagoge*
	Vesalius, *De humani corporis fabrica*, ed. 2.
1556	Juan Valverde de Hamusco, *Historia de la composicion del cuerpo humano*
1558	Realdo Colombo, *De re anatomica*
1561	Gabriele Falloppia, *Observationes anatomicae*
1564	Vesalius. *Anatomicarum Gabrielis Falloppii observationum examen*
1564	Death of Vesalius

List of Figures

CHAPTER 1

Introduction

In 1566, the court physician and astrologer to the emperor Rudolf II, Thaddeus Hagecius (1526/7–1600), compared the achievement of Andreas Vesalius in anatomy to that of the astronomer Copernicus, whose declaration that the earth revolved around the sun, and not the reverse, was published in the very same year, 1543, as Vesalius' *De Humani Corporis Fabrica*.[1] The ideas of these two men and their followers, he declared, had fundamentally changed one's view of mankind.[2] He was not wrong. Vesalius remains the most famous name in the history of anatomy, and his book is now a treasured possession of major libraries across the world. It is seen as a landmark in the history of printing, and its size and beauty still astonish all who see it.[3] The quincentenary of its author's birth in 1514 occasioned conferences and exhibitions from Toronto to Kazakhstan, as well as learned books and articles. Yet, paradoxically, the timing of these celebrations was unlucky, for many of the subsequent publications were already partially outdated through discoveries made from 2010 onwards. Much more is now known about the life and background of Vesalius (with which this study is only secondarily concerned) and, not least, his work as a physician, which occupied most of the last twenty-one years of his life. More of his correspondence has come to light, as well as new details of his family and friends.

V. Nutton, *Andreas Vesalius and his Fabrica, 1537–1564*, Palgrave Studies in Medieval and Early Modern Medicine, https://doi.org/10.1007/978-3-031-69565-0_1

Most striking of all has been the discovery of two sets of annotations by Vesalius himself that were preserved in private collections. The existence of the first, written into his copy of the 1538 edition of the *Institutiones anatomicae (Principles of Anatomy)* that bears both his name and that of his teacher, Johann Guinter, became widely known only in 1991 when the catalogue of the Haskell Norman Collection was published. Its contents were further described in the 1999 sales catalogue of the Collection and in the catalogue of the Toronto exhibition of Vesaliana of 2014 when it was on public display. But it was not until 2017 that a complete transcription and translation of the notes was published along with an English translation of the *Principles.* Five years earlier, in 2012, a long article had summarised an even more remarkable resurrection, the extensive notes that Vesalius had written into his own copy of the second, 1555, edition of the *Fabrica.*[4] The most important were translated in the footnotes to the 2014 English translation of the *Fabrica* by Dan Garrison and Malcolm Hast, which was fittingly published in Basle by the firm of Karger, but a complete transcription and translation of the notes is still lacking. This would be a Herculean task that would also risk being misleading, for many of Vesalius' markings take the form of underlinings or proof corrections that do not change the meaning of the text. Together, these two sets of notes take us into the mind of Vesalius, and, along with printed marginalia in a neglected work, the second edition of his *Paraphrasis,* they allow us to write a history of the *Fabrica* from the moment when Vesalius decided to abandon his plan for a further revision of the *Institutiones,* if not even earlier from his students days, until the publication of his final book, a reply to one of his critics, shortly before his death in 1564. Such an opportunity to follow in detail over a long period the creation and development of one of the most important of all pre-modern medical and scientific works is extremely rare—William Harvey and Isaac Newton are parallel instances, but from the next century—and I have been privileged to be able to assist in this process.

This study concentrates on the development of the book. It resembles an archaeological investigation in that it not only looks back at Vesalius' early ideas on the anatomy of the human body and his attitude towards his own publications, but it also considers the materials involved in the wider context of the new print culture. It thus expands on the conclusions of Martin Steinmann and Sachiko Kusukawa about the significance of Oporinus in the printing of illustrated medical and botanical books in the Renaissance.[5] In doing so it draws attention also to unfamiliar works of Vesalius such as the revised edition of the *Paraphrasis* and the second edition of the *Fabrica,* which has been unjustly neglected compared with its predecessor. This account differs also from the recent survey of the 1543 *Fabrica* by Sachiko Kusukawa by considering its history down to the present day, and its transformation from an academic monograph to a cultural icon world-wide.[6]

Writing at length about the *Fabrica* also presents an opportunity and a challenge. The standard biography of Vesalius, by C. Donald O'Malley, was published sixty years ago in 1964 to coincide with the 450th anniversary of Vesalius' birth and the 400th of his death.[7] Excellent for its time, it stands in need of considerable revision to take account of new discoveries. The celebrations fifty years later produced many valuable articles in books and journals, but no major study of Vesalius' life and times has since appeared on the scale of O'Malley's. This book is not that biography, and hence many of the details of Vesalius' life have been deliberately omitted, nor is it a study of Vesalius' abilities as a surgeon and anatomist, which would require a very different book from someone familiar with the difficulties of cutting into a human body. Rather, it tries to show how Vesalius came to write his most famous book, beginning with his studies in Paris and his early publications in Louvain and Basle, and how he revised and corrected it over two decades. The final chapter continues the story of the book down to the present. In considering the *Fabrica* principally as an object, and looking at the processes whereby Vesalius' words, ideas and visual images were turned into print, it also brings together new material that adds to our understanding of a remarkable and multi-talented man in a period when both anatomy and the world of printing were changing rapidly. Vesalius was at the forefront of innovation in both, although the *Fabrica*'s status among modern anatomists has tended to obscure how unusual he was in his exploitation of the power of print, as well as playing down the roles of his Basle printers, Robert Winter and Johannes Oporinus. Particularly over the last twenty years, as I have worked closely on Vesalius' writings and, especially, his manuscript annotations, I have come to appreciate more and more what he was trying to do and the extent to which he succeeded. At the same time, I have not attempted to disguise the, often unfair, rhetoric of his polemics against Galen and his even more deceitful silences about his debts to others, the source of much contemporary criticism. Finally, by examining closely the way in which the book and its related publications were produced, investigating the medium as well as the message, I hope to show why the *Fabrica* deserves its reputation as the greatest and most influential of all medical books.

Notes

1. Theodorus Hagecius, "Letter to the Reader," in Laurentius Gryllus, *De sapore* (Prague: G. Melantrich, 1566), sig. δ ii r.
2. He was echoing the view of his friend and professor of anatomy at Ingolstadt, Adam Landau (d. 1573), who placed Vesalius at the head of those whose human anatomy had within twenty years completely replaced the animal anatomy of Galen, ibid., sig β ii verso.
3. Elisabeth Eisenstein, *The Printing Press as an Agent of* Change (Cambridge: Cambridge University Press, 1979), used Vesalius as a prime example of the importance of printing as an agent of change, despite more than seventy years gap between it and the beginnings of printing; see Anthony Grafton, "The Importance of Being Printed," *The Journal of Interdisciplinary History* 11 (1980), 265–86.
4. Vivian Nutton. *Johann Guinter and Andreas Vesalius, Principles of Anatomy According to the Opinion of Galen* (London and New York: Routledge, 2017); Vivian Nutton, "Vesalius Revised: his Annotations to the 1555 *Fabrica*," *Medical History* 56 (2012a), 415–43.
5. Martin Steinmann, *Johannes Oporinus, ein Basler Buchdrucker um die Mitte des 16. Jahrhunderts* (Basle and Stuttgart: Helbing & Lichtenhahn, 1967); Sachiko Kusukawa, *Picturing the Book of Nature. Image, Text and Argument in Sixteenth-Century Human Anatomy and Medical Biology* (Chicago: Chicago University Press, 2012).
6. Sachiko Kusukawa, *Andreas Vesalius. Anatomy and the World of Books* (London: Reaktion Books, 2024).
7. C. Donald O'Malley, *Andreas Vesalius of Brussels (1514–1564)* (Berkeley and London: University of California Press, 1964); Stephen N. Joffe, *Andreas Vesalius, the Making, the Madman and the Myth* (Bloomington: AuthorHouse, 2014) is a much shorter account by a distinguished Vesalian.

CHAPTER 2

Before the *Fabrica*

The Young Vesalius

Andreas Vesalius was born to be a courtier. He came from a medical family from Wesel on the Lower Rhine, hence his Latin name Vesalius, which had moved to Brussels, the capital of the Duchy of Burgundy, when that region was at its height of prosperity in the early fifteenth century.[1] His great-great-grandfather, Peter (or, more likely, Everard) Witing, so he believed, had written a commentary on the Arabic physician, Avicenna, before briefly attending the emperor Frederick at some point during his long reign from 1440 to 1493.[2] His son, John (c. 1405–1476), was a leading member of the faculty of medicine at the newly founded university of Louvain from around 1430 before leaving twenty years or so later to become one of the civic physicians at Brussels. Soon after 1450, he dedicated a treatise on plague to Francesco Visconti, Duke of Milan, at whose university of Pavia he had earlier studied for his medical doctorate.[3] As well as maintaining property in Wesel throughout his life, John developed close relations with the Burgundy court, treating Duke Charles of Burgundy (reigned 1467–1476) and possibly also Mary of Burgundy, the future wife of emperor Maximilian I.[4] His son, also called Everard, continued the family tradition by studying at Louvain and becoming a physician to members of the imperial family, but he died in 1484/5 before his father.[5] It was his books and manuscripts, probably inherited from his

V. Nutton, *Andreas Vesalius and his Fabrica, 1537–1564*, Palgrave Studies in Medieval and Early Modern Medicine, https://doi.org/10.1007/978-3-031-69565-0_2

own father, that Vesalius read and used for his own studies fifty years later. They included commentaries on some of the Hippocratic *Aphorisms* as well as on the treatises of the Arabic physician Rhazes that he consulted for his own first publication in 1537.[6] His son, Andreas, was born illegitimate but prospered enough to be trained as an apothecary and to marry Isabel Crabbe, the daughter of a tax-official and city councillor. It is not clear when he was appointed a court apothecary, a post less well paid than that of the doctor but bringing in far more than the income of a normal physician. It was a job that frequently entailed wearisome journeys away from home accompanying the ruler to Spain and elsewhere. Vesalius was thus familiar from childhood with two of the essentials for court employment: stamina and a knowledge of how to behave in the presence of the very great. Another essential, gentility, was aided by a grant of legitimation to his father in 1531 in recognition of his long and dutiful service.[7]

The young Andreas was born, the second of four children, on the last day of December 1514, if one can trust a much later horoscope.[8] His early education was typical of that of a wealthy young man in his region. In 1529, he left Brussels to study at the Castle School of the University of Louvain, where he embarked on the arts course.[9] He studied rhetoric, philosophy and logic at a university that, for a brief moment, was at the very forefront of the new humanist movement in education in Northern Europe. Its new Trilingual College, whose lectures were open to all in the university, offered instruction not only in Latin, the staple of all medieval and renaissance elite education, but also in Classical Greek and Hebrew. Like the greatest of all Netherlands humanists, Erasmus of Rotterdam (1436–1566), its professors wished to inculcate in their pupils an understanding of the foundations of scripture, in Hebrew and Greek, and of the classical authors of Greece and Rome. They aimed to revive a style of writing in Latin that avoided what they saw as the linguistic solecisms of the Middle Ages in favour of the more elegant Latin of the ancient Romans such as the orator Cicero or the historian Livy. They stressed the need for a variety of expressions for the same activity or object, complex sentences that would encompass a series of interlinked statements in an architectural whole, and a flowing rhetoric that would add an elegant literary polish to any theme. Vesalius' Latin is a tribute to his teachers, although his delight in variation, the long periodic structure of his sentences and paragraphs, and his orotundity are not always to the taste of modern readers, accustomed to a more straightforward manner of expression.[10] It is also an ideal medium for disguise, offering an apparent abundance of information while suppressing any details that might occlude the self-image of the author. Its message is Janus-faced, describing events in the present but utilising quotations and allusions from the distant past. It presents new ideas and a challenging message while simultaneously drawing on the centuries-old authority of the ancient classics.

While at Louvain, if not earlier, Vesalius turned his mind to medicine. He may well have been conscious of his family traditions, but, apart from a section in the *China Root Letter* and a passing reference in the second edition of the *Paraphrasis*, he says nothing about them.[11] Nonetheless, it was a move that was certainly encouraged by a family friend, the imperial physician Nicholas Florenas, who had studied both at Louvain and at Bologna. In the preface to his 1537 *Paraphrase of the Ninth Book of the* Almansor, Vesalius credits Florenas with setting out the list of books that he should study, and it was probably he who suggested, or at least supported, his decision to move to Paris in 1533.[12] It was a wise move. Louvain was a typical small medical school, conservative in its ways at a time when the face of medicine was changing rapidly. Its outlook was very much local, training physicians who would live and practise in the Netherlands and Northern France. If Louvain had a wider reputation, it was for the production of annual calendars forecasting the weather and future events for the coming year.[13] This was something signalled out by Hubert Barland (fl. 1510–50), when introducing his 1529 edition of the letters of Giovanni Manardo to the young students of Louvain, although he himself never obtained a teaching post there but instead gained a probably more lucrative one as a court physician elsewhere in the Netherlands.[14]

Paris and Humanist Medicine

By contrast with Louvain, Paris had long been the leading medical school North of the Alps, both in the number of its students, although it was extremely small by modern standards, and in the reputation of its professors. The Paris printers, such as Simon de Colines and the Estiennes, were keen to publish the very latest works on medicine, and the presence of the French court also drew to the metropolis doctors and surgeons from all over France and beyond. Relations between the University, the physicians, the surgeons and the city's barber-surgeons were not always harmonious, especially as the physicians wished to impose their writ and their privileges on all other medical groups, but there was also co-operation. Simply being in a city the size of Paris allowed Vesalius to see what was going on in chemists' shops and to observe how physicians and surgeons worked with a variety of patients.[15] Teaching took the form of lectures on some earlier texts in Latin translation, especially of the Greeks Hippocrates and Galen, and the Arabs Avicenna and Rhazes, in which the professor expounded the meaning of these, often obscure, treatises and emphasised the principles of medical theory and practice that they contained. The Paris faculty library was filled with manuscripts of these much earlier authors, for their long survival within a tradition that went back to the very origins of Western medicine in classical Greece was thought to guarantee the effectiveness of their precepts. Practical experience, by contrast, was briefer, gained by working with a master once one had grasped the theoretical principles that governed health and the workings of the body.[16]

If the Paris medical curriculum was similar to that of Louvain, the way in which it was taught there in the early 1530s was very different. While elsewhere in Northern Europe, medical teaching remained heavily focused on medieval Latin translations or treatises, the Paris faculty was deeply involved in what is often called humanist medicine. This was still based on these earlier authors, but its proponents insisted on a revision or a clarification of the principles behind them through a return to "purer sources", the words of the Greek and Latin authors on whom subsequent authors largely depended. Beginning in the late fourteenth century in Italy, this conviction of the superiority of classical learning in literature and philosophy over what had been studied in the Middle Ages was widely shared around Western Europe by the 1480s. It produced new ideas of proper behaviour as well as a new sense of style in both language and art, and no person with any pretensions to learning could reject it.[17]

But initially at least, physicians and, still more, the less well-educated surgeons stood a little apart from this renaissance. Although classical texts of literature, philosophy and history were quickly rediscovered, and from the 1460s onwards were made more accessible through the new medium of printing, this was not so in medicine. Only one previously unknown Latin medical text from Antiquity was made available before the 1480s, the treatise *On Medicine* by Aulus Cornelius Celsus, fl. 40 CE, which was found in Siena in 1426 and printed in 1468. But Celsus was not himself a doctor but a learned gentleman, whose encyclopaedia of *Artes* aimed to educate others of his ilk in practical skills that might become useful to them. His subjects included estate management, rhetoric and military science as well as eight books devoted to medicine and surgery. Written in an elegant Latin and beginning with a sketch of the whole history of medicine from Hippocrates to his own day, it was a valuable compendium of information that the wealthy reader, like Celsus himself, might put to good use in looking after his household and friends. But this was not the same as practising medicine for money and it depended above all on a sound knowledge of the individual patient in good health as well as bad, something that Celsus thought unlikely in large (in his day slave) hospitals. How Celsus' book was used in the fifteenth century is still far from clear, but, together with the medical sections in the huge *Natural History* of the Elder Pliny (d. 79 CE), it provided a new and much more 'classical' Latin vocabulary that the modern physician could employ in his writing.[18]

Nonetheless, much of the medicine of antiquity long remained hidden from those who called themselves humanists, for one simple reason: most of the ancient medicine that had survived the centuries had been written in Greek. Greek was rarely studied at school until the early years of the sixteenth century and access to Greek medical texts was limited even among those who could read them, for Greek manuscripts were hard to find even in Italy. The Turkish conquest of Constantinople in 1453 had brought about a change, for many Greek refugees fled to Venice, bringing Greek texts with them. Wealthy patrons like the Medici in Florence or the Greek Cardinal Bessarion, who had strong links with Venice, set about collecting manuscripts for their libraries, sending envoys to the Aegean region to seek out long-forgotten treatises as well as employing their own copyists.[19] Crucially important in this search for ancient Greek medicine was the long-lived Niccolò Leoniceno (1428–1524).[20] Born and educated in Vicenza and with a prodigious reputation for his linguistic skills, he took a medical degree at Padua. Of his visit to Northern Europe in 1453, we know little except that he visited Oxford with an Englishman he had met in Padua and was forced to drink wine by the over-exuberant clients of a Dutch bar. He may have taught medicine and philosophy for a while at Padua before moving to the university of Ferrara, where, save for occasional periods of teaching in Bologna, he stayed until his death. He was so much an institution there that it was said that winter began when he put on a cap, and summer when he removed it.[21]

His home university of Ferrara, ever since its foundation in 1391, had been a pioneer of the new humanist learning, in deliberate contrast to the more traditional Padua and Bologna.[22] The Este rulers were intent on increasing their reputation as patrons of culture and made their city and its university one of the major intellectual centres of Europe for more than a century and a half. Leoniceno and his passion for the language, literature and medicine of the ancient Greeks fitted perfectly into this new society, attracting students from N. Italy and beyond. As well as his university teaching, he was paid from time to time as a ducal physician, and between 1470 and 1490 he was occupied in translating for Duke Ercole many of the ancient historians, particularly from the Greek, as well as the humorous *Dialogues* of Lucian and the modern *Roma Instaurata* of Flavio Biondo. Some of these he turned into Latin, but many also into Italian, most of them, alas, now lost or preserved only in manuscript. He was on friendly terms with leading humanist scholars such as Urceus Codrus in Bologna, Giovanni Pico della Mirandola and Politian in Florence (whose ruler Lorenzo de' Medici made tentative approaches to obtain his services as a household physician), and, above all, with the printer Aldus Manutius and his circle of Hellenists in Venice.[23]

By 1490, Leoniceno had acquired the largest collection of Greek medical and scientific writings in all of Europe. At his death, he owned at least 75 Greek codices, many composed of multiple treatises by a variety of authors and encompassing works by almost all the classical medical writers known today.[24] He made good use of them in his own writings. Speaking in 1491 in a debate before the court at Ferrara, he set out a novel manifesto that, when it was published by a local printer the next year, caused a sensation. Its title, *De Plinii et aliorum in medicina erroribus*, *The Mistakes of Pliny and Others in Medicine*, gave only a slight hint of its power to subvert, for this was a wholesale attack on the basis of all Latin medicine from Late Antiquity onwards. In it, Leoniceno argued that the Latin texts that every modern doctor studied were filled with innumerable errors because of the complex process of the transmission of the sources on which they drew. The multitude of translators, the different routes these earlier writings had taken, and the variety of languages involved, Syriac, Arabic, Hebrew and several varieties of Latin, to say nothing of the vagaries of copyists and the misunderstandings of readers, had led to errors of all kinds. To take but two examples, the names of plants and diseases were confused and misunderstood. Pliny had thought '*leucographis*' was a plant, although the Greeks Dioscorides,

Galen and Paul had described it as a mineral, and he had mistaken the herb '*struthius*' for its homonym, a bird. The major Arabic authors, Avicenna and the pharmacologist Sarapion, were taken similarly to task for errors whose consequences could be seen in the fallible productions of those who confected pills and other remedies, as Leoniceno did not fail to point out as he walked the streets of Ferrara.[25] These mistakes were not simply the fault of medieval scribes: even the Roman Pliny had made many errors in transliterating plant names from the Greek botanists into Latin. Only by returning to the original Greek of Galen and the botanist Dioscorides could one be sure of the actual meaning of the words that they had used and, consequently, of the validity of the principles that depended on them and had been inculcated for so long.

There were many others who followed Leoniceno's example, collecting Greek manuscripts and translating them into Latin, including the Englishman Thomas Linacre (1460–1524), physician and tutor in the English royal household and one of the founders of the London College of Physicians in 1518.[26] These 'hellenisers' argued that many of the texts of medieval medicine, and especially those taken from the Arabic, were filled with errors that derived from a misunderstanding or miscopying of the original words of the Greeks, especially Galen and Hippocrates. If only one could return to the original sources, understand and translate them correctly, and even add to their number by bringing to light treatises that had remained for centuries unnoticed among Greek manuscripts, then one would have a much clearer and sounder basis for the principles and practice of medicine. Confusion would be swept away through philological exactitude, and one would indeed have brought about a rebirth of medicine. These new medical standards, enforced throughout medicine by universities and colleges of graduate physicians, would, it was argued, ultimately benefit all mankind.[27]

These claims, widely trumpeted in Italy, were accepted only slowly North of the Alps, for two related reasons. They depended on a knowledge of ancient Greek, something rarely studied in schools or universities even within Italy before 1500, and, equally important, on access to Greek books and, particularly, Greek manuscripts. Only one ancient Greek medical writer, the botanist-pharmacologist Dioscorides, had been printed in full, in 1499, and only two of the many writings of the most prolific of all ancient writers, Galen, had been published in their original Greek, *The Method of Healing* and the smaller *Method of Healing, for Glaucon*, in 1500. Neither appears to have circulated widely outside

Northern Italy. The situation was a little different when translations into humanist Latin began to be published at the end of the century. Printers played safe, publishing modern translations direct from the Greek of a handful of basic texts that were read by all medical students, and often printing them alongside the older versions. (This was not foolish, for earlier medieval writers, and not a few contemporary lecturers, used the older translations and cross-checking with the new versions in a separate publication was cumbersome at best.) As Leonhard Schmaus, a doctor from Salzburg, no German backwater, angrily complained in 1519, it was no use saying that one could gain more solid learning from a day reading Galen than from a whole week of Avicenna, when editions and translations were few and far between, or available in only a few major cities.[28]

By the time Vesalius arrived in Paris in 1533, the situation there had changed dramatically from what it had been only a few years earlier. Paris was now a major centre of the new medical humanism, the preference, wherever possible, for the classical authorities over the medieval, with consequent substantial changes in the texts studied. Throughout the medical faculty, from the lecturer in surgery, Jean Tagault (d. 1545) to its Dean, Jean Vasses (1486–1550), Parisian scholars were eagerly translating, commenting upon and putting to wider use the newly rediscovered riches of the Ancients. The arrival in print of the works of Galen in Greek in 1525, of Hippocrates in 1526, and of several other Greek writers in the next decade, led to a flood of new translations into Latin, many published for the first time by De Colines, and later and occasionally into French, and to a reorganisation of the texts chosen for comment. The Faculty was among the first purchasers of the Aldine Greek edition, and its members swiftly made good use of what the new Galen had to offer.[29] Not only did they discover a less dogmatic Galen, whose uncertainties offered the possibility of new interpretations of familiar material, but his philosophical writings and his Hippocratic commentaries, for the most part entirely unknown before in the Latin world, offered new models for thinking about medicine, most evident in the writings of Jean Fernel (1497–1558).[30] Others, notably the two Faculty members who most influenced Vesalius, Jean Dubois (Sylvius) and Johann Guinter, were heavily involved in the dissemination and systematisation of the new Galenic medicine through the medium of lectures and translations.

Sylvius (1478–1555) was the older man. Trained in the classical languages, as well as in Hebrew and mathematics, he came to medicine with a strongly philological bent. But he had a practical side to him that is often forgotten. He had spent some years travelling to learn botany and pharmacology, subjects revived by the humanists under the influence of Galen and Dioscorides. He had gained his medical degrees relatively late in life, taking his MD at Montpellier in 1530, before moving to Paris where he taught medicine at the Collège de Tréguier while he waited for the Paris Faculty to grant him a licence to practise and teach there. It was not until January 1536 that, along with Fernel, he was given the right to teach in the Faculty, so that Vesalius' earliest mentions of' "my learned teacher" and a scholar who "can never be praised enough", will refer to the few months that followed as well as to his earlier teaching at the Tréguier. His first publications dealt with fevers, dietetics and pharmacology, in which he emphasised the importance of following a method (a favourite word among the medical humanists) on the model of Galen.[31] This method could best be gained by studying these works from the past in a new order (although one based on Galen) that showed how all the parts of medicine fitted together, the subject of one of Sylvius' early and most influential publications.[32] In his later treatise on anatomy (finished in 1542 but not published until 1555), he combined a Galenic method of dissection and some of his own discoveries with an exposition of the anatomy of man in Galenic terms, although he was aware that Galen had largely dissected animals.[33] A later pupil in the 1540s refers to him refusing to let others dissect for him and bringing in under his robes a mouldy arm or thigh of a dissected criminal to illustrate Galen's meaning.[34] At least until 1542 Vesalius always spoke of him with respect, and his recollections of his teaching do not depict the ferocious Galenist of his last years.[35]

Guinter (1497–1574), a native of Andernach on the Rhine, was the rising star of the Faculty. He had travelled widely, including a period at Leipzig studying medicine, before he came to Paris in 1527, first teaching at the Collège de Cornouaille before joining the Faculty in 1536. He had made his reputation as a phenomenal translator of Galen—a joke had him translating many lines of Greek into Latin while standing on one leg—and of other Greek medical authors.[36] At this period in his life, he was a vigorous and outspoken Galenist, attacking his colleagues for their ignorance of the true meaning of Galen, despite their claims to be his adherents.

The dissection of a corpse as part of medical education was a relatively recent and infrequent innovation, beginning in Italy in the 1310s.

The first such dissection in Paris may not have taken place until the end of the fifteenth century, and in most Northern universities not until the middle of the next century.[37] But, everywhere, all would-be physicians learned about the structures and workings of the human body primarily from books, whose emphasis was less on the anatomical details of each part of the body than on its function and purpose within the wider scheme of God's creation. This was a message that had been delivered frequently and at length by Galen and by his medieval interpreters.[38] Chief among them was Mondino de' Liuzzi (c. 1270–1326), a Bolognese professor who around 1316 persuaded the university authorities to allow him to dissect a corpse in front of students, and whose treatise on anatomy, his *Anothomia*, quickly became the standard source of information about the body.[39] When a formal dissection did take place as part of the medical course, the body was cut up by a surgeon or an assistant, the *sector*, while someone else, the *expositor*, expounded the words of Mondino (which were often read and clarified by a third person) and attempted to set what could be seen and was being described into the broader context of medical and philosophical knowledge. Learning how to dissect, regarded as manual or 'mechanical' work, was left to aspiring surgeons. Most medical students were content with an occasional view of a corpse that largely confirmed the truths they had learned in earlier lectures.[40]

Mondino's *Anothomia* remained the basis for anatomical teaching throughout European universities well into the 1530s, for medieval scholars knew only a handful of minor or truncated Galenic texts on the subject and thus had only a very partial understanding of the role dissection had played in his medicine as a whole. It was the 1525 publication of the Aldine Greek edition of Galen's three major treatises on anatomy that first revealed to a wider audience who could read Greek his conviction that anatomical knowledge gained through dissection lay at the very foundation of the medical art. Galen had aimed the three treatises at different audiences. The previously unknown *On the Opinions of Hippocrates and Plato* explained the body in Platonic terms, *On the Usefulness of Parts*, fuller and more accurate than its medieval Latin abridgement, *De juvamentis membrorum*, stressed the functions or purposes of the parts of the body, their teleology, as Aristotle put it, while the third, *Anatomical Procedures*, although preserved only in part, was by far the most detailed and accurate account of how to dissect the body yet available. Together, they quickly made dissection the slogan of the new humanist medicine.[41]

Nowhere was this change more swift or more apparent than in Paris. In the early 1520s anatomical study at Paris had been largely peripheral to student interest. as a series of printed eulogies of medical graduands in 1523 shows. Most of them are praised for their skill in diagnosis and in understanding medical theory, a few are commended for the surgical knowledge they had gained from Professor Tagault's lectures and two or three are singled out for their expertise in medical astrology. Of anatomy and dissection, there is not a word.[42] But a decade later, it was all the rage. Students, and not just medical students, clambered over walls and trampled over gardens to secure front seats at a lecture, while many members of the Faculty chose to lecture on it or discuss it in print.[43] Guinter's 1531 translation of Galen's *Anatomical Procedures* made Galen's most important practical treatise on anatomy accessible to a wide audience effectively for the first time, and easily outsold the rival (and slightly earlier) translation of the work by Demetrius Chalcondylas and Berengario da Carpi.[44] Vesalius attended Sylvius' course of lectures on Galen's *On the Usefulness of Parts* in 1535, although he later claimed to have found them jejune, passing over large sections and omitting most of the information on how to dissect as too difficult for his audience.[45] The following year, he acted as dissector for Guinter's public lectures on anatomy as well as assisting him in his private courses (of which we know next to nothing).[46] In one of the formal anatomies, he had the chance to dissect a prostitute who had been hanged in the prime of life, something unusual because male bodies for dissection massively outnumbered female, and the statutory obligation in many universities to dissect a woman every three years was often avoided, sometimes for decades, because of a shortage of suitable bodies.[47] He was chosen to help probably because his interest in dissection was already well-known to his teachers. He mentions "long hours" spent in the Cemetery of the Innocents turning over graves in a hunt for bones, sometimes alone but often with friends, who might bet with one another that they could identify a bone blindfold.[48] (An insistence on the feel and texture of individual bones can be found throughout the various versions of the *Fabrica*.) It was also dangerous, for "many savage dogs" prowled around the site of the public gibbet at Montfaucon, where one might have the chance to inspect a whole corpse of an executed criminal.[49]

Other Parisian teachers were also hard at work on anatomy, and one should view sceptically Vesalius' later description in the *Fabrica* of official lectures as "a few casual and superficial demonstrations of a few organs by unskilled barbers".[50] Tagault, as the main teacher of surgery, often alludes to his own expertise as a dissector in his 1543 book on surgery, and two physicians were also engaged in writing their own books.[51] The Spaniard Andrès Laguna (1511?–1560), who took his bachelor's degree in medicine in 1532–3, stayed on in Paris for a few more years, publishing his small *Anatomica methodus*, *Anatomical Method*, there in 1535. It is an elegant summary of the new humanist anatomy, with many literary flourishes and citations of Greek, but, although Laguna occasionally refers to his and Tagault's experiences with a corpse, the treatise largely restates older ideas in a new stylistic form.[52]

Of greater ambition was a project contemplated and begun in the early 1530s by Charles Estienne, a member of a distinguished French family of scholars and publishers, assisted from an early stage by the surgeon Estienne de la Rivière. Estienne, who had also studied medicine at Padua, aimed to produce an anatomical treatise including many illustrations on a scale and of a complexity unknown before.[53] Untangling the complex history of his book is far from easy, for both text and illustrations have their own genealogies, and one cannot assume that the printed version of 1545 corresponded in every respect to what was originally intended either in content or in audience.[54] What is clear is that most of the work for it was carried out in Paris, to which Estienne had returned in 1532, during the period when Vesalius was also there.[55] Four woodcuts, produced in the workshop of Geoffrey Tory and François Jollat, bear dates between 1530 and 1532, and most of the rest, up to the middle of Book 3, were certainly set up in print by 1539.[56] It incorporated new discoveries and had the majority of the illustrations in the later books made from actual dissections performed by the surgeon Estienne de la Rivière, who was assisting at anatomies in Paris between 1535 and 1538 and who may have been involved in the choice of illustrations from the very beginning[57] (Fig. 3.2). Co-operation between teacher and surgeon did not run smooth. In 1539, de la Rivière instituted a lawsuit to ensure that he gained appropriate recognition for his role in the book. It delayed publication for years, but there is nothing to suggest that, when his book finally appeared, Estienne had taken the opportunity to make use of the new information in the *Fabrica* that had appeared shortly before. He does, however, seem to have used the first three plates of Vesalius' *Sex Tabulae* in modified form, almost certainly via the publications put out by Christian Wechel in 1538–9.[58] What, if anything, Vesalius owed to Estienne is hard to detect. He can have known both Estienne and de la

Rivière at best only briefly and there are no obvious signs of their influence. The first, and almost certainly the earliest, book of *De dissectione* is a conventional exposition of the new humanist anatomy and lacks any of the illustrations prepared by de la Rivière, and the later books may have only just been begun, if at all, when Vesalius left Paris. But, whether or not he was aware of Estienne's book, impressive though it was when it finally appeared, his own was projected on a much grander scale than any Parisian work on anatomy and used images in a more sophisticated way, as we shall see. Nonetheless, the stir caused by the arrival of the new 'humanist' Galen within and without the Paris Medical Faculty, not only provided an exciting environment for the young medical student but also led very swiftly to a wide recognition of his ability with the knife among his friends and some at least of his teachers.

From Louvain to Basle: The *Paraphrasis*

The outbreak of war between France and the Emperor in July 1536 put an end to Vesalius' stay in Paris, for it was dangerous for someone with connections to the imperial court to remain there.[59] He returned to Louvain, inspired by the new medical humanism, only to find old-fashioned defenders of the indefensible at the head of the medical faculty. Hubert Barland in 1529, encouraged by none other than Erasmus himself, had hoped to introduce the new Parisian ideas there and saw in Manardo's letters something that would be valuable to all doctors, even the supporters of Avicenna, but he succeeded in only stirring up a hornet's nest. Three years later, he accused Arnold Noots, one of the two ordinary professors of medicine, of serious mistakes in his knowledge of simples and in following the "hallucinations" of Avicenna in his practice of bloodletting.[60] Noots' colleague in the second chair of medicine, Leonard Willemaers, may have been less hostile, for Vesalius in 1538 sent him his best wishes as a very talented follower of Galen, but both men were later forced to resign after having been accused of living in the wrong century.[61] Noots' successor, the younger Jeremiah Drivère (Thriverus, 1504–1554), was scarcely more sympathetic to many of the views of Paris and the Italians, defending in his first book, in 1532, traditional dietetics in cases of arthritis against "the dreadful misunderstandings of the neoterics".[62] He and Vesalius later clashed in a very public disputation over the proper site for bloodletting, with Drivère calling the leading humanists "the Lutherans of medicine".[63]

If Vesalius' relations with some of the Louvain doctors were difficult, his time there was not wasted, for he was able to widen his experience in anatomy. It may have been through the support of Johannes Armenterianus (d. 1560), another professor of medicine, that he was allowed to lecture on the structure of the human body. Although as yet he had not obtained a medical degree, he was also permitted to perform the official anatomy under the chairmanship of Armenterianus, whose anatomical knowledge he appreciated.[64] He later claimed that this demonstration had benefitted the students greatly, for no such anatomy had even been contemplated for eighteen years, as well as improving his own expertise beyond what he had been able to show in Paris.[65] Through his contacts with men of wealth and influence he was allowed to attend, and even participate in, autopsies, including that of a young girl of eighteen in the service of the Countess of Egmont. On this occasion, left behind after the departure of the family and the physician who had carried out the autopsy, in his view incompetently, he took the chance to do a little dissection himself to examine the hymen.[66] When he sought to obtain bodies for his own private dissections, his family connections came into play, for he was protected by the friendly burgomaster. He had no qualms later about proudly recording how he stole a body hanging from a gibbet outside the town gate, and prepared and mounted it as a demonstration skeleton, supplying the missing hand and leg bones from another corpse so cleverly that he could pass it off as something he had acquired earlier in Paris.[67]

It was at Louvain in early 1537 that Vesalius published his first book, an updating, a *Paraphrasis*, or *Paraphrase*, of Book IX of the *Liber ad Almansorem*, a large medical handbook dedicated to the ruler al-Mansūr of Khorasan by the Arab physician Rhazes around 900 and, in the Latin translation by Gerard of Cremona, one of the staples of university teaching.[68] Book IX offered a brief guide to the treatment of diseases from head to toe, based mainly on Galen's writings, and was but one part of a much bigger work in ten books. Although he later embarked on a much more careful reworking of the other nine books, there is no indication at this stage that he had this larger project in mind.[69] As he explained in his preface and in his address to the reader at the end of the volume, this was not a translation, but a paraphrase, an attempt to provide for this medieval author the sort of revision that Leoniceno and his peers had achieved for other standard texts that had derived from Greek Antiquity.[70] There was one major difference from what they had done: Vesalius did not know Arabic, so he could not have returned to the original, but relied solely upon Gerard's Latin. The book bears all the marks of his Parisian education. He turned somewhat hispid medieval Latin into modern flowing sentences and introduced headings and subheadings in the margin, as well as an index of chapters, to help guide the reader.[71] Occasionally he added an original Greek word in the margin or explained that the word he had chosen in the text was his own interpretation or was now the standard term, superseding that of the earlier translator.[72] This was not always easy, for some of the ingredients mentioned had not been known in antiquity or were, like 'blitum', which worked like spinach, unfamiliar: no wonder that some physicians had opted for words that were downright stupid or consequently employed a remedy for angina that was extremely dangerous.[73] Like the new translators of Galen, in order to decide what a passage meant, he compared the different versions of the Latin text he found to see if any changes were needed. Some copies, he noted, recommended three weeks rest instead of three days when recovering from being almost frozen to death, while other versions of the same recipe had different ingredients.[74] He compared statements and recipes in other medieval writers such as Avicenna and Mesue, and often in their ultimate ancient sources.[75] He relied mainly on Galen, Paul of Aegina and Aetius, but he mentions Theophrastus, Alexander of Tralles and, from Latin authors, Celsus, Pliny and Scribonius Largus.[76] Among modern authors, he cites only Leoniceno and Manardo, whose *Medical Letters*, in particular, had showed him how traditional Arabo-Latin medicine could be improved by a return to the Greek, either by clarifying the meaning of some Arabic words in Latin or by showing how certain medical practices had wrongly departed from their Classical Greek origins.[77] The practical merits of this short handbook were further enhanced by a few marginal notes criticising or developing what Rhazes had said on the basis of what Vesalius and his contemporaries now did or found in modern pharmacies.[78] The longest of them explains the difference between Galen's views on the appropriate site for letting blood and that of the Arabs, the subject of his disagreement with Drivère.[79] But there are also some interesting sidelights. Some pastilles are imported directly from the East, and Vesalius notes that 'modern' Greeks practise a different form of cautery from the Arabs.[80] On the title page, he is described as a 'candidate in medicine', which suggests that most of the book had been written, or most material collected for it, while he was a student in Paris before returning to

Louvain. It is thus an important witness to how much of the new medical humanism he had absorbed while there.

The *Paraphrasis* was printed at Louvain by Rutger Rescius, a teacher of Greek at the university who dabbled as a publisher. Vesalius' dedicatory letter from Brussels to his friend and sponsor Nicholas Florenas is dated February 1, 1537, the book's title page gives simply February 1537, and the letter to the 'fair-minded reader' at the end is dated March 1537. But this is not the whole story. The copy of the *Paraphrasis* in the library of the former Paris Faculty contains instead of the letter to the reader a letter from Vesalius in Louvain to a clerical friend Rutger Giffeus simply dated to 1537 (Fig. 2.1). Giffeus, originally from Bois-le Duc (s' Hertogenbosch), had been a friend from his early days in Louvain, but he had chosen to study theology, staying on for at least a couple of years after graduation as a lecturer there.[81] The Paris letter to Giffeus is unique, for in all the other surviving copies, it is replaced by the address to the reader, which differs in its wording only at the beginning and end (Fig. 2.2). In both versions, the placement of the letter is odd, coming at the end of the book but with sentiments that would better suit a preface. Vesalius may well have planned to include a letter to the reader from the start, as Maurits Biesbrouck has suggested, but his further suggestion, that the letter to Giffeus was printed as a gift to his friend after the original printing had been completed is unlikely, for there can be no doubt that it was the letter to Giffeus that was set up in print first. In the Paris copy its text neatly fills the whole of the second side of the leaf, sig. O iii verso, whereas in the others the final sentence of the slightly shorter letter to the reader is stretched out in an ugly fashion to avoid a large blank space. No sane printer, even when pressed to help a friend, would have undertaken to print a single copy that would, at the very least, have involved a substantial resetting of existing text and casting new type. If the Giffeus letter came first, it would have taken a skilled setter an hour at best to make the necessary alterations before the relevant leaf could be pulled again, for he could easily have reused existing letters. Although the whole of the text, including the final two folios of the *Paraphrasis* itself, would already have been set up and much even printed by the end of February, making the index, setting and printing the remaining sheets and checking for errata would all have taken time. There is no reason, then, to reject the date at the end of the letter to the reader that places the publication of the whole book in (probably early) March 1537. It was only later, after the letter to Giffeus had been printed, that someone, perhaps even

Vesalius himself, noticed that its first word had been printed incorrectly as 'SYNCERERISSIMO' and deleted the extra letters by hand.[82]

Between the book's publication and Vesalius' examination for his degree at Padua in December 1537, we have no assured date for any event in his life. The colophon of the second edition of the *Paraphrasis* declares that it was printed at Basle by Robert Winter in March 1537, a date that has been universally accepted, even if with reservations. In this scenario, a revised version of the book would have been delivered to Basle within days of being printed, and printing there would have begun straight away in order for the book to appear in April: the colophon would then refer to the date printing began, not when it ended. This is a tight schedule that, as will be shown, is impossible for many reasons, including the length of the journey itself. In September 1518, the distinguished humanist Erasmus took almost three weeks to reach Louvain from Basle, travelling by boat down the Rhine to Strasbourg and then overland by

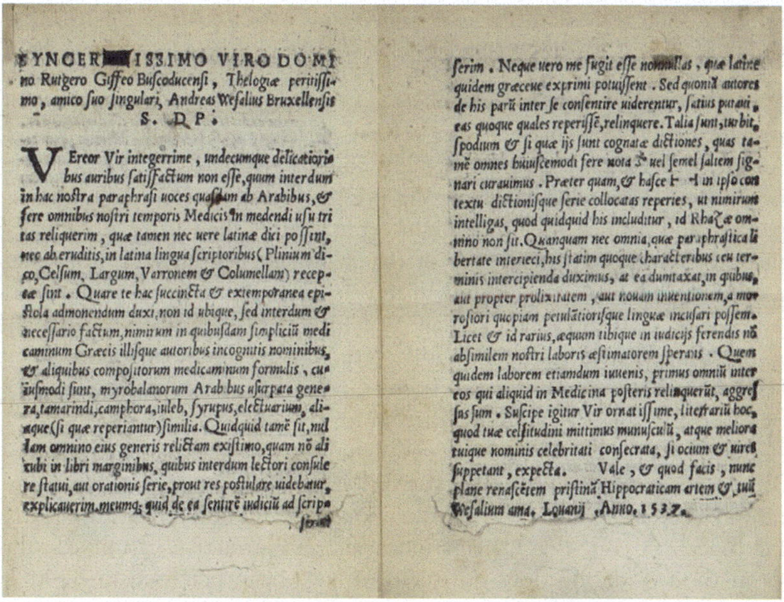

SYNCER[redacted]ISSIMO VIRO DOMINO Rutgero Giffeo Buscoducensi, Thelogiae peritissimo, amico suo singulari, Andreas Wesalius Bruxellensis S. D. P.

VEreor Vir integerrime, undecumque delicatioribus auribus satisfactum non esse, quum interdum in hac nostra paraphrasi uoces quasdam ab Arabibus, & fere omnibus nostri temporis Medicis in medendi usu tritas reliquerim, quae tamen nec uere latinae dici possint, nec ab eruditis, in latina lingua scriptoribus (Plinium dico, Celsum, Largum, Varronem & Columellam) receptae sint. Quare te hac succincta & extemporanea epistola admonendum duxi, non id ubique, sed interdum & necessario factum, nimirum in quibusdam simpliciū medicaminum Graecis illisque autoribus incognitis nominibus, & aliquibus compositorum medicaminum formulis, cuiusmodi sunt, myrobalanorum Arabibus usurpata genera, tamarindi, camphora, iuleb, syrupus, electuarium, aliaque (si quae reperiantur) similia. Quidquid tamē sit, nullam omnino eius generis relictam existimo, quam nō alicubi in libri marginibus, quibus interdum lectori consulere statui, aut orationis serie, prout res postulare uidebatur, explicauerim, meumq; quid de ea sentirē iudiciū ad scripserim. Neque uero me fugit esse nonnullas, quae latine quidem graecene exprimi potuissent. Sed quoniā autores de his parū inter se consentire uiderentur, satius putaui, eas quoque quales reperissē, relinquere. Talia sunt, turbit, spodium & si quae ijs sunt cognatae dictiones, quas tamē omnes huiuscemodi fere nota ✱ uel semel saltem signari curauimus. Praeter quam, & hasce ⊢ ⊣ in ipso contextu dictionisque serie collocatas reperies, ut nimirum intelligas, quod quidquid his includitur, id Rhazae omnino non sit. Quanquam nec omnia, quae paraphrastica libertate interieci, his statim quoque characteribus seu terminis intercipienda duximus, at ea dumtaxat, in quibus, aut propter prolixitatem, aut nouam inuentionem, a morosiori quopiam petulātiorisque linguae incusari possem. Licet & id rarius, aequum tibique in iudicijs ferendis nō absimilem nostri laboris aestimatorem speraus. Quem quidem laborem etiamdum iuuenis, primus omniū inter eos qui aliquid in Medicina posteris reliquerūt, aggressus sum. Suscipe igitur Vir ornatissime, literariū hoc, quod tuae celsitudini mittimus munusculū, atque meliora tuique nominis celebritati consecrata, si ocium & uires suppetant, expecta. Vale, & quod facis, nunc plane renascētem pristinā Hippocraticam artem & tuū Wesalium ama. Louanij, Anno. 1537.

Fig. 2.1 Andreas Vesalius, *Paraphrasis in nonum librum…Rhazae*, Louvain: R. Rescius, 1537, letter to Giffeus. (Copyright: Bibliothèque Nationale de France)

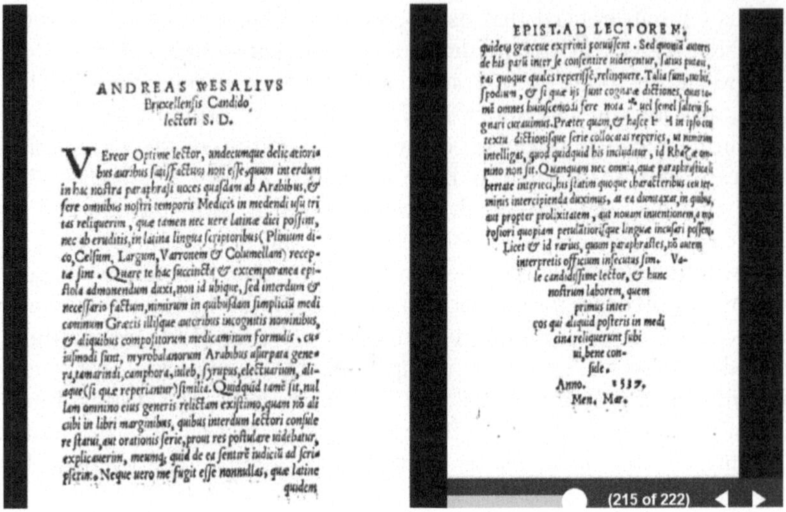

ANDREAS VESALIVS
Bruxellensis Candido,
lectori S. D.

VEreor Optime lector, andecumque delicatioribus auribus satisfactus non esse, quum interdum in hac nostra paraphrasi uoces quasdam ab Arabibus, & fere omnibus nostri temporis Medicis in medendi usu tritas reliquerim, quæ tamen nec uere latinæ dici possint, nec ab eruditis, in latina lingua scriptoribus (Plinium dico, Celsum, Largum, Varronem & Columellam) receptæ sint. Quare te hac succincta & extemporanea epistola admonendum duxi, non id ubique, sed interdum & necessario factum, nimirum in quibusdam simplicium medicaminum Græcis illisque autoribus incognitis nominibus, & aliquibus compositorum medicaminum formulis, cuiusmodi sunt, myrobalanorum Arabibus usurpata genera, tamarindi, camphora, inleb, syrupus, electuarium, aliaque (si quæ reperiantur) similia. Quidquid tamē sit, nullam omnino eius generis relictam existimo, quam nō alicubi in libri marginibus, quibus interdum lectori consulere statui, aut orationis serie, prout res postulare uidebatur, explicauerim, meumq; quid de ea sentirē iudiciū adscripserim. Neque uero me fugit esse nonnullas, quæ latine quidem

EPIST. AD LECTOREM.

quidem græceue exprimi potuissent. Sed quoniā autores de his parū inter se consentire uiderentur, satius putaui, eas quoque quales reperissē, relinquere. Talia sunt, nabit, spodium, & si quæ ijs sunt cognatæ dictiones, quas tamē omnes huiuscemodi fere nota * uel semel saltem signari curauimus. Præter quam, & hasce ⊢ ⊣ in ipso etiam textu dictionisque serie collocatas reperies, ut nimirum intelligas, quod quidquid his includitur, id Rhazæ omnino non sit. Quanquam nec omnia, quæ paraphrastica libertate interieci, his statim quoque characteribus seu terminis intercipienda duximus, at ea duntaxat, in quibus, aut propter prolixitatem, aut nouam inuentionem, a morosiori quopiam petulātiorisque linguæ incusari possem. Licet & id rarius, quum paraphrastes, nō autem interpretis officium insecutus sim. Vale candidissime lector, & hunc nostrum laborem, quem primus inter eos qui aliquid posteris in medicina reliquerunt subiui, bene consule.
Anno. 1537.
Men. Mar.

(215 of 222)

Fig. 2.2 Andreas Vesalius, *Paraphrasis in nonum librum…Rhazae*, Louvain: R. Rescius, 1537, letter to the Candid Reader. (*Source* Wellcome Collection)

carriage and on horseback.[83] The reverse journey, upstream, would have taken a little longer. Using a speedy courier to deliver a copy of the printed book immediately it left the press, a possibility O'Malley leaves open, would reduce the time slightly but would then leave no opportunity for Vesalius to make any of the many corrections, additions and expansions that are to be found in the notes in the margin of the revision.[84] Nor can one imagine that Vesalius set off himself immediately for Basle and Padua, rereading and correcting his book on the journey like a London commuter on a train. Given this short time-scale, it is not surprising that Cushing included in his bio-bibliography of Vesalius a suggestion (which he did not accept) by his Yale colleague Leon Nemoy proposed that Winter was using the older form of dating that began the year in late March, and thus that the book itself did not appear until early March 1538.[85] But this explanation, although attractive, is unlikely for this method of dating does not seem to have been practised in Basle at this time. Indeed, Balthasar Lasius and Thomas Platter, Winter's partners, had used the newer form of dating, March 1537, for a volume that cannot have been published in 1538 because it is presupposed by another of their publications dated to March 1538.[86] The colophon date must thus be an error, and its redating opens up a new chapter in our understanding of Vesalius' life.

The second edition of the *Paraphrase* is the least studied of all Vesalius' writings. The great nineteenth-century biographer of Vesalius, Moritz Roth, mentioned it only in a footnote and his use of it as the source for his citations of the first edition indicates that he regarded the two as being little different.[87] More recently, Maurits Biesbrouck and Omer Steeno have provided a succinct English introduction to the whole work, detailing the contents and the subsequent printings of this 'enigma', although they accept the traditional date of publication and assume a general similarity between the two editions.[88] They at least studied its contents, unlike O'Malley, who merely noted its publication (and subsequent reprints) and its title page. He dismissed its contents curtly. "At any rate, by this time the work ought rightly to have been forgotten and probably would have been had not Vesalius later acquired fame".[89] But this was not the view of Vesalius, who saw it as a valuable way of reconciling the theoretical and cultural claims of humanist medicine with the achievements of the Arabs and medieval writers in practical medicine. Others agreed with him, for there were later editions published at Lyon in 1552 and at Wittenberg in 1586, and Rhazes, Avicenna and Arabic pharmacology continued to be studied for several decades to come.

O'Malley seems to have followed Roth in believing that this second edition added nothing of importance. That is largely true of the main text. Winter corrected almost all the typographical errors in the Louvain edition that were signalled in the Errata (and added a few of his own), but the differences between the two editions of the text are not significant, being mainly stylistic. The biggest change occurs at p. 36, lines 5–16 = 1537a, 21r, lines 9–15, where a single long sentence was divided, and Vesalius clarified what he had previously written by adding a few words at the end of the sentence. The Latin style of the sentence is also much improved. In the beginning 'modo ante' replaces 'Paulo ante', and the cumbersome use of gerundives to indicate what should be done is abandoned in favour of the more elegant "you will examine" or "you will assist the patient in this way".

This tinkering with the wording to improve the style is a constant feature of Vesalius' activities as an author and need not detain us further here.[90] But if he made only relatively minor changes to his text, he made around 150 additions and alterations to the marginalia, many of which reveal more of his earlier life in Paris and Flanders.[91] Rescius' weaknesses as a printer are here obvious. His typesetting is poor with many letters squashed together and set in a font that makes consultation difficult. In his revision, Vesalius expanded wherever possible the abbreviated titles and chapter and page references to ancient and medieval authors and introduced more headings to guide the reader. Some of these may have been omitted unintentionally by Rescius, but others point to new information that Vesalius thought valuable. He gives new references to five ancient authors, Dioscorides, Galen, Aetius, Paul and Archigenes, the last probably taken over via Galen.[92] He adds more references to the Arabic pharmacologist Mesue and also to Johannes Arculanus (fl.1450), whose popular commentary on the *Ad Almansorem* by Rhazes may be the source for some of the new comments in the margin.[93] He is interested in variant readings, whether in the printed editions of Arabic authors, or in the manuscript of the *Ad Almansorem* belonging to his grandfather which lacked the relevant words found in the earlier printings.[94]

The rest of the annotations reveal Vesalius the practical physician. He remarks on the differences between Greek and Arabic practice, and sometimes on different views in medieval authors, especially in relation to bleeding or in the provision of alternative constituents of remedies.[95] He comments on the names given to specific drugs or herbs in contemporary pharmacies (*officinae*) and sometimes warns the reader that a particular therapy is now useful only in particular circumstances, superseded or even to be avoided.[96] Much of this he will have picked up during his stay in Paris, but at least one new reference relates to his native region: in Brabant, there are some bushes that bear red fruits (blackberries?) that are useful in cases of vomiting.[97] Together they show how much he had learned, and put into practice, before he had reached Italy and confirm that, even at this stage, he was far more than an anatomist.[98]

To write up these annotations, as well as to set them up in print, would have taken a considerable time, perhaps several months, and a hypothesis that they were handed over in Basle during Vesalius' travel to Padua perhaps in August or September in time for the beginning of term on October 18, 1537, is plausible. It would also help to resolve two related questions that, as far as I know, have never previously been posed: why should Vesalius have chosen Winter as his publisher? and why should Winter have agreed to publish the revised version of a book printed elsewhere only a few months earlier? Robert Winter was not an obvious choice as a publisher, nor was a small book by a young and relatively unknown Fleming something to be eagerly welcomed as a potential best-seller. The answers to these questions throw new light on an obscure period in the life of Vesalius and of the *Fabrica*.

It is important to remember that in early 1537 neither Basle as a publishing centre nor Vesalius as a writer had achieved the eminence that they were to enjoy in 1543, and, perhaps yet more important, Winter and Vesalius had never met. Particularly if Vesalius' revised version had been sent by courier (and it would have taken considerable time for Vesalius to make his preparations to leave for Padua along with servants and a chest or chests of clothes and books, if he delivered it himself), there was no sensible reason why any printer should have chosen to publish a book by an unknown author that had appeared in a roughly similar form only matter of months before. Vesalius was a wealthy man who might be expected to make a financial contribution to the publication costs, but Winter could not count on this. Besides, what he was being asked to do was something unusual and extremely risky: publishing a revised edition of a book only months after its first appearance and before he could be sure that it had already sold well.

Although corrected and revised editions of theological and literary works by now were common, this was not so in medicine.[99] Best sellers, particularly translations of the medical classics, might be republished without the author's consent in a different place where copyright protections did not apply, and with only a few changes, apart from the straightforwardly typographical. Vernacular self-help books, like the treatise on distillation ascribed to Michael Puff von Schrick, Heinrich Stromer's *Treatise on Plague*, or Eucharius Roesslin's book on gynaecology 'for pregnant women and midwives', his *Rosegarten*, found a ready readership in German, and Gregor Reisch's large Latin compendium of information on a variety of subjects, including medicine, his *Margarita philosophica*, went through several allegedly improved and updated editions within eight years of its first appearance in 1504. But books intended for a more academic market, or at least for one that could read and understand Latin, were reprinted in new editions only rarely until the 1540s or even 1550s. More often, as with the printings of the collection of basic university texts such as the *Articella*, or with Manardo's *Medical Letters*, updates added blocks of new material rather than carrying out even a cursory revision of what was already in print. But improved editions were not entirely unknown. Berengario da Carpi's *Isagoge*, his *Introduction to Anatomy*, first published in 1522, was republished less than a year later with significant new illustrations, probably reflecting some of his recent dissections.[100] But this was a short introductory university text, intended to be bought by a constantly changing constituency. Books meant for the wider medical community were revised less often.

Nonetheless, two younger scholars, both later familiar to Vesalius, had published revised editions of their own small compendia of diseases and therapy in the years preceding the *Paraphrasis*. The much-travelled Janus Cornarius (1500–1558) brought out his *Sketch of Medicine* with Froben in Basle in 1529 and a revised edition five years later with the same press.[101] Leonhard Fuchs of Tübingen (1501–1566) published his *Introduction to Medicine* first at Haguenau in 1531 and in a second edition in nearby Strasbourg four years later.[102] Like Vesalius, both authors provided a short guide to medicine that introduced the new humanist practical medicine to a wider Latinate community, especially local physicians. Winter may well have known Cornarius' book, but there is no guarantee that this example of a revised version was known to Vesalius. Significantly, even when new editions of any books were published there was often a gap of years between them. Vesalius, by contrast, was correcting his *Paraphrasis* from the moment it left the press, and the initiative for publishing may have come from him rather than from Winter, although one should not believe that he had intended to publish with him in Basle from the start.[103]

Until the mid-1530s. Basle, like Louvain, had catered for a largely local readership of medical and scientific texts. But as Frank Hieronymus has shown, from then on Basle printers turned from publishing in German to rival Paris and Lyon in their coverage of Northern Europe for Latin books.[104] Basle's situation on the Rhine and at the junction of trade routes to France, Italy and Northern Europe gave it easy access to many markets, and the reputation of Cratander, Froben and Thomas Platter had been spread widely by humanists such as Erasmus as well as by theologians. At the centre of a growing network of publishers, Basle was far better suited to the plans of the ambitious Vesalius than other publishing centres North of the Alps, let alone Louvain, especially as, for the moment, publication in Paris was tricky for political reasons.

But Winter was not an obvious choice for someone living hundreds of miles away as he was only just starting out as a printer and publisher. Two years earlier, in 1535, the more experienced Lasius and Platter had formed a printing business with Robert Winter and his brother-in-law,

Johann Herbst, better known as Oporinus (Fig. 2.3). How much Oporinus contributed to the partnership at this point is uncertain, apart from his linguistic skills, for he was an ambitious man with many other projects, being at various times headmaster of the St Leonard's school (where he secured a post for the talented but relatively poor Thomas Platter) and an assistant to Paracelsus.[105] Lasius and Winter both came from the upper echelons of Basle society, being members of prestigious local Guilds, and Winter, who had married a goldsmith's daughter, was wealthy. But, whatever their personal qualities, they were both poor businessmen, often teetering on the brink of bankruptcy, and the partnership was soon in difficulties. Lasius, although technically a good printer, seems to have been a difficult character, and by spring 1537, if not a little earlier, Winter was branching out on his own. Indeed, Vesalius' book was among the first to be published under his own name.[106] Vesalius may have known of the firm through his old teacher, Johann Guinter, for Lasius and Platter had published his *Institutiones anatomicae, Principles of Anatomy*, the previous year, and two of his Galenic translations, under the editorship of Sebastian Sinckeler, were to appear under their imprint in March 1537.[107] But in neither volume is there any mention of Robert Winter.

All this suggests that there had been little or no contact between Vesalius and Winter before the middle of 1537, and that, now the March date for the second edition has been eliminated, a much more plausible scenario can be constructed. Once Vesalius had decided, like many other sons of rich families, to move to Italy and Padua to study for a higher medical degree, his easiest route was via Basle. He would thus avoid France and be able to spend time in one of the citadels of Northern humanism, Basle, following in the steps of Hubert Barland. Even if he did not bring with him a recommendation from Guinter, Rescius or his cousin Martin Stern, who may have already moved there, his wealth and his humanist background at Louvain would have made him a welcome visitor. Basle was a small city, able to be crossed in any direction in little more than a quarter of an hour, and with strong connections between the university, its school and its many publishers. Once introduced into this close-knit community, Vesalius will have found it easy to make friends. Whereas we have no information of when he came to Basle or how long he stayed there, we can be sure from their subsequent relationship that he was soon on the closest of terms with Winter and Oporinus. In August 1543, on his second stay in Basle, Vesalius stood as godparent to Winter's son Hans Rudolf, and, later, tried unsuccessfully to introduce Oporinus'

Fig. 2.3 Johannes Herbst (Oporinus), (1507–1568), publisher and printer of the *Fabrica*. Paul Freyer, *Theatrum virorum eruditiorun clarorum*, Nuremberg: J. Hoffmann, 1688, p. 1460. (*Source* Wellcome Collection)

wastrel stepson Jacob to court life.[108] It is not fanciful to imagine that he and Winter discussed some of his plans together and that it was as a result of this personal engagement that Winter offered or agreed to publish his revision.[109] A corrected version would have been handed over before Vesalius left, although when it appeared in print must still remain unclear.[110] It was still a risk, mitigated by their personal acquaintance, but it was a decision with enormous consequences for both men.

Humanist Anatomy Before the *Fabrica*

In his decision to go to study in Padua Vesalius was following a well-trodden path[111] Wealthy students from all over Europe, from England, Spain, Germany and Poland, made their way there to hear some of the great names of the day and to obtain the most up to date information on everything from anatomy and medical botany to the best methods of examining patients or of dealing with the contagion of plague. The 1530s, after a period of stagnation and economic uncertainty, saw a marked upswing in the fortunes of the university, encouraged by a Venetian government eager to profit economically from visitors to its university. The recently introduced rule that no local citizen could hold a chair not only prevented the formation of dull dynasties of inbred teachers but also allowed the authorities to bring in new talent and ideas from elsewhere.[112] Although the authorities at Bologna might disagree, Padua could now claim to be the leading medical school in Europe.[113]

Vesalius had left Paris without taking his degree, although his decision to call himself a 'candidate in medicine' on the title page of the first edition of *Paraphrase* asserted that he had done a good deal of the work required for one. Although he may have contemplated taking a medical qualification at Louvain (and, given the gaps in the degree records, he may have done so), it is more likely that he decided that there was enough evidence in the volume to show his practical experience and chosen instead to aim for a medical doctorate from another and more prestigious university, Padua.[114] His studies in Paris and Louvain, even without taking a degree, will have satisfied the academic requirements for a Paduan MD, while a period of treating the sick in Venice under the supervision or sponsorship of "the most famous physicians there", together with the evidence of his new book, was enough to allow him to be examined for his doctorate in the first days of December.[115] On December 6, immediately after his final examination and graduation, the young Vesalius was appointed to a lectureship in surgery and anatomy at a beginner's salary of 40 florins. Whereas in N. Europe surgery was rarely taught as a university subject, in Italy, it formed a small part of the medical course, usually alongside anatomy, and would-be surgeons could gain some of their training at the university rather than entirely by surgical apprenticeship. Why Vesalius was chosen is not clear—the swiftness of his appointment is not unusual—but he probably already had a reputation for his anatomical skills since his time in Paris and impressed the faculty during the few weeks when he was in Padua preparing for his MD examination. It was a momentous appointment. He began teaching straight away, giving lectures on surgical inflammation and dissecting his first corpse on December 6.

Anatomical dissections had for some time formed a regular part of medical education at Bologna and Padua, where the, at least annual, demonstration lasted for over three weeks. It was held in winter and followed the order established two centuries earlier by Mondino, beginning with the abdominal cavity, thorax, head and brain and ending with the extremities and bones. It was a solemn, formal occasion. After the appropriate religious rites, the body of a recently executed criminal was brought into the place of dissection, where a temporary stage and seating would be erected, often in a church, an appropriately chilly venue for dealing with a corpse before it could putrefy. (The first permanent anatomy theatre, that of Padua, was not erected until 1594–5 after Vesalius had departed.)[116] If the evidence of woodcut illustrations is to be believed, there were three main participants, the *sector*, usually a surgeon, who did the actual cutting, a *lector* in full academic dress, who read out from his high stall the relevant passages in Mondino's *Anatomy*, and a third man, the *ostensor*, probably the senior professor, who pointed to the specific part mentioned in the text and put what was being shown into an appropriate context.[117] In private dissections, like those of Guinter in Paris, the actual cutting was done by an assistant, either a surgeon or a student, while the teacher explained what might be seen. At the close of the proceedings, the dissected portions of the body were collected together and given an appropriate religious burial. Animals were also dissected, usually dogs, cats or pigs, to supplement (and, since obtaining human bodies was always problematic, often stand for) what could be shown in the human specimen, and they could, as Galen had argued, be vivisected to show the body in action.

Mondino had relied for the anatomical details of the body upon Galen's *De juvamentis membrorum*, supplemented by the anatomical sections in two minor tracts, *De motibus dubiis*, *Problematical Movements*, and *De voce et anhelitu*, *Voice and Respiration*. These three works often circulated together in medieval manuscripts to form a small anatomical corpus, but their focus was less on how to cut up a body than on showing how a knowledge of the body could resolve certain problems as well as reveal the teleological artistry of the Creator. By the end of the fifteenth century, all three were becoming suspect as more of Galen's anatomical writing became available in the original Greek, and their deficiencies were revealed.[118]

The initial response to the arrival of Greek manuscripts of Galen's anatomical writings in Italy was primarily linguistic, as has already been noted, above, p. 11 and focussed on the clarification of terminology. The new information was first introduced to a wide readership in 1501 by Giorgio Valla (1447–1500) in a huge encyclopaedia with an uninformative title, *De expetendis et fugiendis rebus*, *Things to Look for and Avoid.* Its nine books on medicine, 24–30 and 47–48. were based principally on Celsus and two late Greek excerpters of Galen, Aetius of Amida and Paul of Aegina, which Valla had read in manuscripts in his own library.[119] Later printers, particularly in Germany, broke his large work up into separate and more manageable sections, each with their own title. That dealing with anatomy, given the title *De humani corporis partibus, The Parts of the Human Body*, was reprinted several times, the last as late as 1585, and provided the general reader with an insight into the new medicine and its more accurate terminology.[120] It was in the following year, 1502, that Hellenised anatomy was first used in a medical textbook. Its author, Alessandro Benedetti (1452–1512), had travelled in the Greek world and owned a few Greek manuscripts himself. Provocatively he gave his work a purely Greek sub-title in transcription, *Anatomice*, a direct challenge to the earlier tracts with very unclassical Latin titles such as *Anothomia.*[121] The friend of leading humanists such as Merula and Ermolao Barbaro, Benedetti had begun his work in Venice in the 1480s, making use of Galen but also of the lexicographical information provided by ancient writers such as Celsus, Rufus of Ephesus and the lexicographer Pollux. His *Historia corporis humani sive Anatomice, An Enquiry into the Human Body or Anatomy*, was an elegantly written and beautifully printed survey, disguising what was a traditional Galenist exposition under a new vocabulary. It came under immediate attack from Gabriele de Zerbi (1445–1505), the leading anatomist at Padua, who countered the claims made for the new terminology by stressing his own considerable expertise with the knife, which provided the essential practical information for the surgeons who attended his course.[122] Improvements in terminology, in his view, contributed little to the understanding of the body compared with knowing how and where to cut. The very appearance of his much larger book, printed in 1502 in double columns in the older Gothic script with many abbreviations, contrasted with the modern roman lettering of Benedetti. The detailed record of his many observations was a deliberate reproach to the Hellenists, few of whom were themselves expert dissectors. Like Antonio Benivieni (c. 1440–1502), whose *De abditis nonnullis ac mirandis morborum et sanationum causis, Some Remarkable Hidden Causes of Diseases and Cures*, recorded his experiences with the knife in both autopsies and treatments, Zerbi had long been involved with all types of surgery as well as presiding over the annual anatomy at Padua.[123] His murder at the hands of brigands in modern Albania called forth various jibes at the manner of his death, portraying it as appropriate revenge for a life spent dissecting dead criminals, and his book was similarly scorned as outdated. It was Benedetti's shorter work that was more often reprinted, even if the Pavia editors of the 1517 reprinting felt it necessary to revert to the older medieval spelling and format familiar from Zerbi.

By far the most significant of the new writers on anatomy was Jacopo Berengario da Carpi (d. 1530), professor at Bologna and the author of two books on anatomy, a short introductory manual and a much longer and more innovative commentary based on his lectures on Mondino.[124] Both were illustrated with far more images than any previous anatomical book. Berengario relied heavily upon the Greek Galen in his criticism of Mondino, almost certainly through his knowledge of the translation by Chalcondylas (1424–1512), which he edited for publication in 1529.[125] But although he cited Galen more than 1500 times and believed that he had rediscovered the path that Galen himself had followed, he was no slavish Galenist. He preached what he termed an anatomy of the senses, *anatomia sensibilis*, encouraging others to see and feel for themselves and to trust in what they could see. He emphasised his own wide experience of cutting open many hundreds of bodies, both human and animal, in public and private, as a surgeon and at formal university anatomies. This had revealed that Galen had made several errors, not least in his description of the veins and in his belief in the existence of a *rete mirabile*, a vascular network at the base of the brain where psychic *pneuma* was produced.[126]

Berengario's substantial commentary on Mondino of 1521 is arguably the best account of the human body before Vesalius' *Fabrica* because of its depth of detail and its constant injunction to follow the author's own example of practical investigation.[127] His much smaller *Isagoge breves ...in anatomiam humani corporis, A Short Introduction to the Anatomy of the Human Body*, of 1522 is clearly written and a major improvement in accuracy over Benedetti despite its brevity. It is also illustrated with a small number of wood cuts, using chiaroscuro to give depth to the images and placing some of the figures in an artistic landscape (Fig. 2.4). It gives greater details of the human body as revealed by dissection than any other student guide, including Vesalius' later *Epitome*.[128] A second edition was published at Bologna the following year and was followed by printings in Strasbourg by Heinrich Sybold in 1530, and again in Venice in 1535 by Bernardino di Vitale, the future publisher of Vesalius' *Tabulae sex*. Given the publishers' claim that it had been issued in response to student demands it is perhaps surprising that Vesalius did not recommend it to his students when he began his teaching in 1537.[129]

The publication of the Aldine Galen in 1525 and, even more, of the subsequent Latin versions of the newly printed anatomical treatises changed everything.[130] Within little more than a decade, anatomical lectures were being delivered to enthusiastic audiences and anatomical textbooks written by eager Galenists around Europe.[131] Matteo Corti (1475–1544), the best-paid medical professor in Europe, expounded Galenic anatomy at Bologna and Pisa while someone else (in 1540 Vesalius himself) dissected the body in front of him and his audience.[132] In Venice, Niccolò Massa was lecturing and demonstrating anatomy in the manner of Berengario, and his own treatise on anatomy, published in 1536, is arguably superior to that of his predecessor. How much Vesalius derived from his book and from attending at least some of his courses is unclear for Vesalius deliberately hides his debts to others. Conversely, Massa, when reviewing the *Fabrica*, could claim that almost all the apparently new discoveries by Vesalius had been public knowledge among contemporary dissectors and that it was only his many obligations, both private and professional, that had prevented him from anticipating the message of the *Fabrica*.[133] At Ferrara, anatomy teaching was in the hands of Antonio Maria Canani (d. after 1551), who was succeeded in 1540 by his kinsman, Giambattista Canani (1515–1579).[134] Others further afield proclaimed the new Galenic message. Around 1538 François Rabelais, whose 1532 edition of the Hippocratic *Aphorisms* contained the first printing of an original Greek medical text North of the Alps, was presiding at a public dissection of a criminal and explaining to an admiring audience in Lyon the wonderful work of the Creator that they could see.[135] At the same time, at the new Lutheran university of Wittenberg, Philip Melanchthon insisted that students from all faculties should

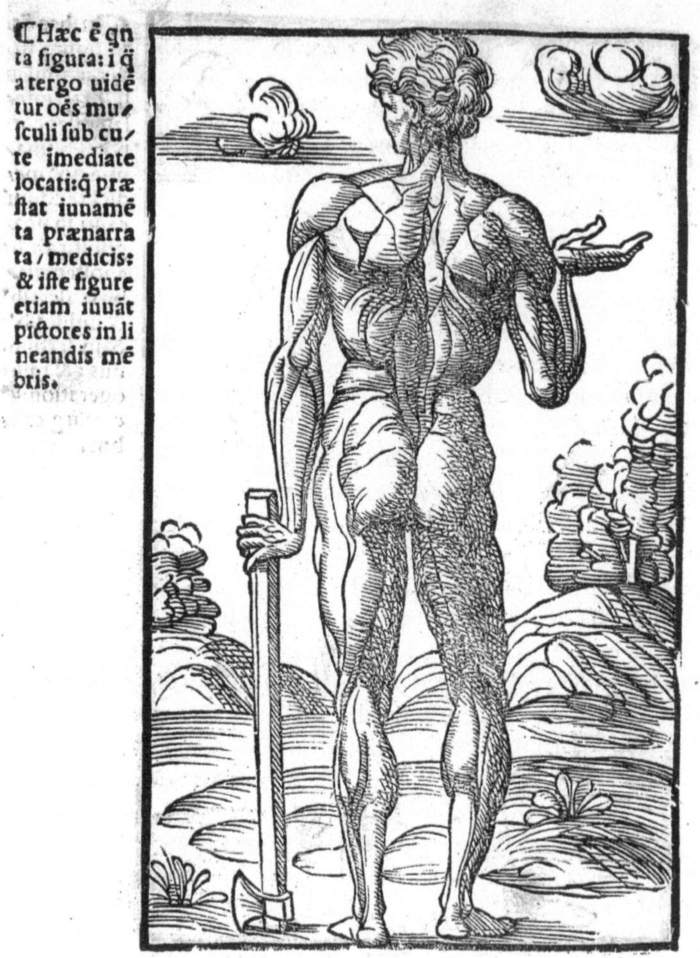

Fig. 2.4 Jacopo Berengario da Carpi, *Commentaria….super Anatomia Mundini*, Bologna: H. de Benedictis, 1521, fol. 520v. The muscles seen from the rear. (*Source* Wellcome Collection)

attend his course of lectures on the soul, half of which were taken up with an account of the human body based on the most up to date research on the new Galen.[136] He was guided in this by his friend, the distinguished Galenist Leonhard Fuchs of Tübingen, who agreed with him about the need to improve on Benedetti's book, which Melanchthon had found skimpy and puerile.[137]

The Young Professor

The extremely short space of time between his degree examination and his new appointment in December 1538 did not allow Vesalius to introduce many novelties into his lectures. One of the student spectators, Vitus Tritonius, recorded that Vesalius in his anatomical presentation was lecturer, demonstrator and dissector in one, and that he recommended his students to follow the anatomy texts of Galen and Guinter rather than Mondino. He also praised Guinter's description of the face muscles, and rather than say more during the lecture itself about the third pair of cerebral nerves, he decided that it would be easier and no less effective to refer his students to what Guinter had written.[138] For his course in 1538, as we shall see, he planned a revised update of his master's handbook.

During his lectures, he had also illustrated some of his findings with his own sketches, some of which were copied down by Tritonius. In his first lecture on surgery, he used a drawing on paper to illustrate Hippocrates and Galen's ideas on bloodletting. This drawing, he later claimed, aroused such interest among both doctors and students that they wanted a drawing of arteries and nerves also.[139] He could not disappoint them, especially as this would be valuable for those attending his lectures on anatomy. But he cautioned that learning about parts of the body and plants from books or formulae alone was hard, if not impossible, although he agreed that visualisation would help to fix things in the memory.[140]

By the beginning of April 1538, his sketches had been turned into a printed form as six anatomical plates, *Tabulae anatomicae sex*, *Six Anatomical Plates*, and published by the same Venetian printer, Bernardino di Vitale, who only three years earlier had republished Berengario's *Short Introduction*[141] (Figs. 2.5 and 2.6). Others might have been sceptical of the value of such aids, not least because of the impossibility of representing accurately a human body on the printed page, but Vesalius was convinced that his future audience would agree that not a line was incorrect.[142] For the accuracy of his statements, even if unexpected, he appealed to the experience of his extremely knowledgeable Parisian teachers, as well as the doctors at Louvain who had attended his dissections there.[143] The *Six Anatomical Plates*, produced in conjunction with the artist Jan Stephan van Calcar, were an immediate success and not only in Padua, but being plagiarised around Europe and establishing new standards of accuracy.[144] Both plates and, to a lesser extent, the new textbook, his revision of Guinter, foreshadow in a limited way what was to come with the publication of the *Fabrica*, fewer than five years later.

The use of sketches in an anatomical or surgical lecture was not in itself new. The surgeon Henri De Mondeville (1260–1320) may have been the first to illustrate his lectures with visual aids, models and drawings, for a picture, no matter how schematic, would often have helped to indicate what was involved in the dissection or how best to cut into a body in an operation.[145] Some earlier Greek manuscripts of surgery had also included images of bandaging, and the luxurious Nicetas codex, copied and illustrated around 940 CE, contained also a series of pictures of surgical procedures. They were designed to illustrate a discussion of the early Hippocratic treatise *On Joints* by Apollonius of Citium (about 100 BCE) and some at least may have been included from the start.[146] Several medical manuscripts in Persian or Arabic contain drawings of a skeleton and of the nervous and vascular systems, accompanying a text that probably went back in some form to late-antique Alexandria, and these may have encouraged the production of similar images in Western Europe from 1300 onwards. The so-called five-picture series (sometimes enlarged to include nine images) depicted the skeleton, the muscles, veins, arteries and nerves, as well as individual organs. They were widely copied, not just in books intended for a medical readership, and are often accompanied by a text that appears to go back to late-Antiquity and, at several removes, to Galen. But they rarely show much resemblance to the human body and are often little more than pretty pictures as in the ‘cubist’ anatomy in Oxford, Bodleian Library, MS Ashmole 399, or the depiction of the nerves in London, Wellcome WMS 290.[147] Some artists attempted to give a three-dimensionality by the use of shading, but they were rare, and the quality of their images much inferior to those depicting other aspects of medical and surgical life. Their aim was far less to show the body and its parts accurately than to provide an aide-mémoire, confirming visually what Galen and others had already described in their books, and to place

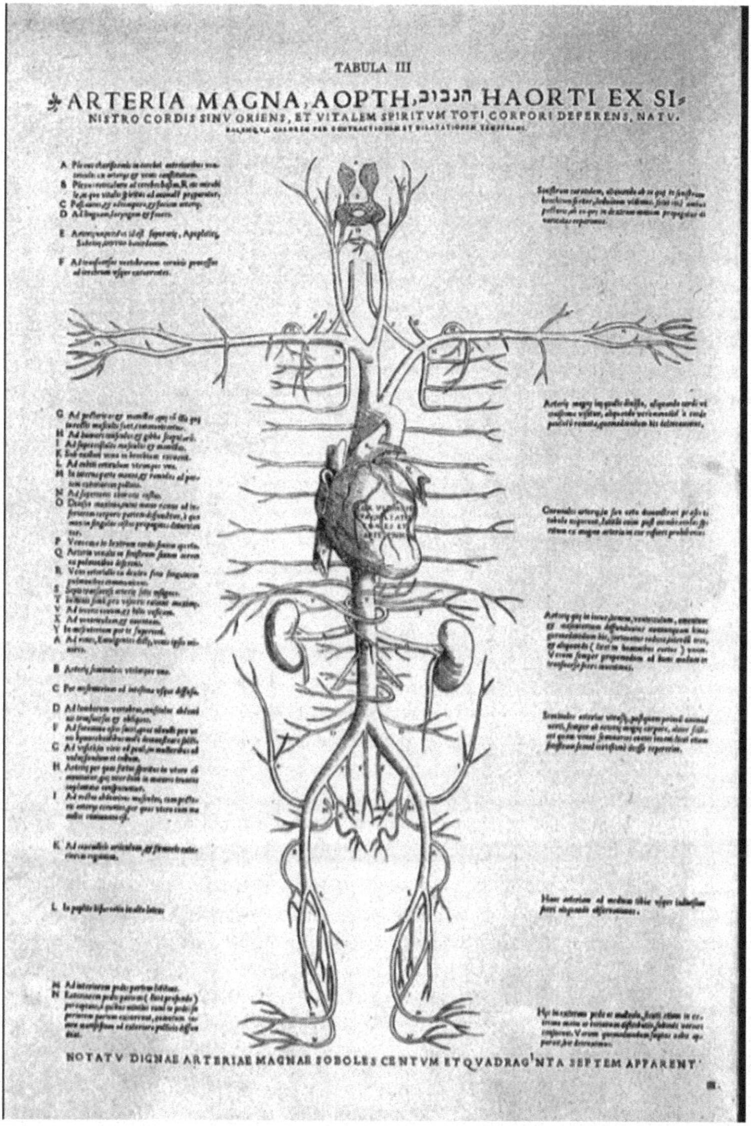

Fig. 2.5 Andreas Vesalius, *Tabulae anatomicae sex*, Venice: B. Vitalis, 1538, pl. 3. The arterial system. (*Source* Wellcome Collection)

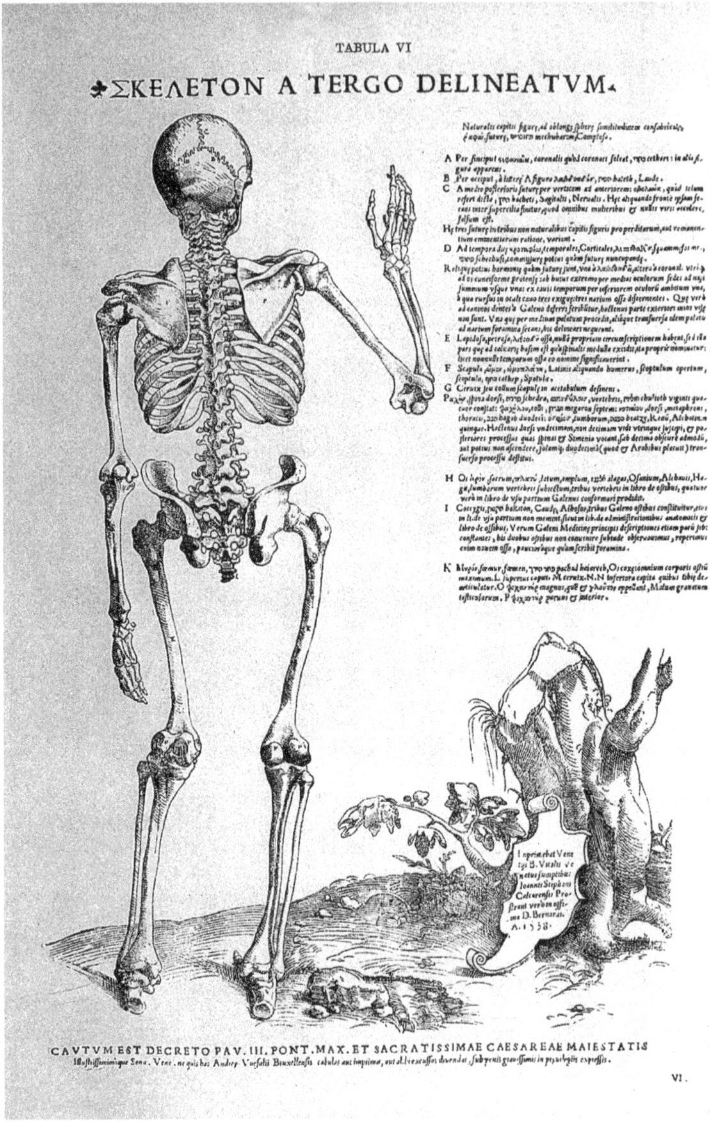

Fig. 2.6 Andreas Vesalius, *Tabulae anatomicae sex*, Venice: B. Vitalis, 1538, pl. 6. The skeleton from the rear. (*Source* Wellcome Collection)

the human body in wider contexts, whether theological, astrological or medical.

Such images continued into the world of print, when they were often termed, confusingly to a modern reader, 'counterfeit'. This does not mean that they were in some way fraudulent, but rather the reverse. They were intended to present individual objects, persons or events "as they were seen", thus emphasising the reality of what was shown rather than the imaginary inventions of artists. This was the title given by an Italian-trained doctor, Wendelin Hoch, to the first of a pair of broadsides put out at Strasbourg in 1517. It showed the opened body of a man who had been dissected there "in the presence of many barbers and surgeons", while a German poem beneath the woodcut described what parts could be seen in this skilful and accurate 'counterfeit'. Around the corpse were set the tongue and representations of the head and brain during the actual process of the anatomy, just as individual parts of the body were depicted around a corpse in some medieval drawings.[148] The second 'counterfeit' was modelled on the tomb of Albrecht von Pfalz-Mosbach, bishop of Strasbourg, d. 1506, in the Collegiate Church of Notre Dame de la Nativité, at Saverne, close to his episcopal palace. It showed a skeleton, with the bones captioned in Latin, with, below, a German poem meditating upon death. Both images were repeated, along with a more traditional 'bloodletting man', by the Strasbourg publisher Johannes Schott in a book on surgery, the *Feldtbuch der Wundarztney* of Hans von Gersdorff, that he advertised at the end of the first poem and was to publish a few weeks later, along with a few more details of the circumstances of the formal anatomy.[149] The original broadside, which was available also in colour, served as an advertisement as well as a source of information and a memorial to two matters of specifically local interest. But, despite Schott's claims for accuracy, all these images resemble more what had been present in earlier manuscripts than a careful delineation of what could be seen in a corpse or skeleton. Indeed, the first image of the dissected corpses has links with an earlier image of the body that Schott had printed in 1503 in the wide-ranging encyclopaedia, *Margarita philosophica*, *The Pearl of Philosophy*, of Gregor Reisch (d. 1523). Miriam Usher Chrisman suggested that this had been done by an artist who had observed a dissection but was still hesitant to make precise drawings. A second version, by Johann Gruninger, is somewhat more exact and may have been improved by an artist who had observed an autopsy.[150] Schott's innovative broadside was intended to provide information for as wide a readership as possible, but this method of publication did not have any immediate successors until Heinrich Vogtherr in 1537 and 1538, nor do his images fully reflect the artistic world in this prosperous region.[151]

Such a disjunction from what contemporary artists and printers were already doing may seem at first surprising. Lifelike statues and paintings were extremely common, and the medical scenes in the various printings of Ketham's *Fasciculus Medicinae*, first published at Venice in 1491, were of a high quality.[152] In Paris, Simone de Colines from 1525 onwards printed illustrations prepared by Geoffroy Tory, and some of his frontispieces are beautiful as well as informative. Bibles and religious texts, and especially Lutheran broadsides, were frequently published with illustrations.[153] Among artists Polluaiolo in the late 1470s clearly relied on dissected parts of human bodies in some of his paintings, and some of the 1528 engravings of the labours of Hercules by the French artist Gabriel Salmon (c. 1485–1542), it is suggested, may contain reminiscences of the illustrations in the anatomical publications of Berengario.[154] Artists of the calibre of Raphael and the young Michelangelo were also drawing dissected limbs, to say nothing of Leonardo da Vinci.[155] There is no longer any doubt that Leonardo collaborated for a few months around 1510 with a young Galenist and medical teacher at Pavia, Marcantonio della Torre (d. 1511), who as the professor in charge of the anatomies there was able to provide Leonardo with bodies.[156] But the precise nature of their collaboration is far from clear, and although Leonardo writes of completing a series of *Dissections*, none of his surviving drawings would easily make an anatomical atlas of sorts, let alone provide pictorial guidance within a course of lectures on Mondino or, more likely, Galen's *On the Usefulness of Parts.* Besides, the extent to which his drawings were accessible after his death is disputed, and stories of his achievements may have had more influence on fellow artists than on medical writers.

It was Berengario da Carpi who was the first anatomical writer to make extensive use of illustration[157] (Fig. 2.4). In his large commentary on Mondino in 1521 he included 21 woodcuts, some of them placed together at the back of the book, and all clearly in the same schematic tradition as earlier images, although more clearly drawn and with greater sophistication.[158] Many of the blocks were reused in his 1522 *Short Introduction* for students, others were recut, and there was a new image of wombs. Several of the figures are artistically placed within an idealised landscape, a plan later followed to a much greater extent by Vesalius and by Estienne. Short summaries explain the importance of the image to physicians, but both drawing and cutting are crude and there is little attempt to link text and image, and the names of the body parts are given within only a few drawings. Medical historians have carefully described the small number of ways in which published images of the body departed from Galenic orthodoxy and, particularly, the extent to which Berengario's images, far more accurate than any predecessor's, can be described as factual illustration.[159] But their findings only confirm that the emphasis of the illustrations was as an aid to memory rather than any accurate depiction of the bones, muscles and vascular system.

One further set of images must be noted, a pair of skeletons that were included first in the Lyon 1535 edition of Balamio's translation of the newly discovered treatise of Galen's *On Bones, De ossibus*. They show a front and rear view of a skeleton placed in a landscape, with a certain amount of shading to give depth and with much greater clarity than those of Berengario. Lind indeed considers their positioning of the skeleton to be more natural than those of the *Tabulae sex*. But they are not linked directly to a text, and they are much smaller than those of Vesalius. They were reprinted by Cratander at Basle in his 1536 edition of Latin translations of Galen, and again in the 1538 Basle edition of a complete Greek Galen, but Vesalius appears to have not known of their existence until after the publication of the *Tabulae*, for he accused both Balamio and his patron, Cardinal Ridolfo, of deliberately withholding this information from him.[160] His failure to mention Cratander's Latin volume in this context is perhaps surprising since he had spent some time in Basle after its publication, and Cratander was a friend of Oporinus. But Vesalius was a master of silence.

The schematic nature of these images of the body prompted Febvre and Martin to conclude that it was in botanical illustration, rather than in anatomical printings, that a desire for realism first manifested itself.[161] Many late-medieval manuscripts depict highly realistic images of plants in herbals and in the margins of books of hours that surpass in both accuracy and quality contemporary depictions of the body in medical texts.[162] Similar images took longer to appear in print, and possibly not until the 1530–36 Strasbourg printing by Johann Schott of Otto Brunfels' *Images of Plants taken from Life, Eicones vivae*, whose illustrations were cut by Johann Weiditz, a prolific Strasbourg illustrator of Biblical and religious texts.[163] The contrast with such earlier crude botanical printings as the 1522 *Erbolario volgare*, printed at Venice by A. de Bindonis, is marked in both style and accuracy and was emphasised by Brunfels in his title, but his illustrations were quickly surpassed by those in the *Enquiry into Plants, De historia stirpium*, of Leonhard Fuchs, published at Basle in 1542 by Michael Isingrin.[164] The deservedly famous portraits of the block-cutter Veit Rudolph Speckle, and of the artists Heinrich Füllmaurer and Albert Meyer at the end of the book that show them drawing and copying a plant placed in a jar before them, rightly acknowledge their major roles in the book's production. Fuchs' images were not only more accurate, better laid out on the page, and of a different aesthetic quality than those of Brunfels but were also correspondingly more expensive. Not everyone accepted Fuchs' innovations, and he provoked debate about the role and validity of illustration. Nonetheless, his book quickly became the model for later books on botany and natural history like those of Pietro Andrea Mattioli and Conrad Gessner, just as the *Fabrica* was to become for anatomy.[165] But it is important to remember that all these publications came after 1538 and that although the existence of Fuchs' book would have been known in Basle months before publication and, as will be argued, influenced the making of the *Fabrica*, Vesalius had already taken the crucial decision to produce an illustrated book on anatomy as early as 1539.[166]

Familiarity with Vesalius and his subsequent publications tends to obscure how unusual his decision was. Berengario's *Short Introduction* of 1522 was the first illustrated anatomical treatise aimed at a student market. Both it and the Venetian reprint were relatively expensive, as were all illustrated books, and the images, taken over mainly from his *Commentary on Mondino* of the previous year, took second place to the text, reinforcing the larger message of the words. In this, it was very different from the *Six Anatomical Plates* that Vesalius issued to accompany his 1538 lectures. Here it was the visual images that dominated, even if they were intended to supplement and strengthen the words that he had written in them and spoken to his audience.

Vesalius may have been involved in a similar enterprise earlier when he was in Paris, for in 1536 the publisher Chrestien Wechel brought out a pair of anatomical sheets, designed for student use and entitled *Osteotome, The Division of the Bones.* No copy of it apparently survives but it was seen in the nineteenth century by Choulant and later by Roth, who described it in detail.[167] Wechel brought out a second edition in 1538, which formed the basis for a careful study by Sten Lindberg in 1953.[168] The 1536 edition would appear to have been the first of its kind: two plates illustrating a skeleton from the front and rear were accompanied both by identifications of the parts and a short exposition for the benefit of students attending an anatomical course or demonstration. Pointing out the striking likeness between the arrangement of the bones here and in the corresponding plate in the *Tabulae sex*, Lindberg concluded that Vesalius must have known of the 1536 printing and borrowed from it, or, indeed, that it was he who had suggested the project to Wechel and supplied him with appropriate material. The latter seems more likely, for Vesalius would have been expecting to continue as Guinter's dissector, had it not been for his sudden return to Louvain, and the 1536 printing would then have served for both the public and the private anatomies with Guinter in which he participated. The fact that Wechel did not bring out a new edition until after the publication of the *Tabulae* may also indicate that the Parisians did not fully appreciate at first what Vesalius had done.

But whatever the extent of his involvement with the *Osteotome*, there can be no doubt that the preparation and publication of the *Tabulae sex* marked a major change in the way in which anatomical images were created for use in a medical context. Save for the *Osteotome* and the earlier broadsides by Schott, above, p. 39, printed images had previously been included within larger books and were consequently much more expensive. Vesalius's six plates, three times the number of Wechel's, constituted a treatise in themselves and were thus available at a price that students and many other non-medics could easily afford. This concern to promote his ideas to a wider readership, evident later in his *Epitomes* of the *Fabrica*, is a notable feature of Vesalius' activity.

Although they depended in part on earlier precedent, the images of the *Tabulae* were far bigger than those of Berengario, Balamio and Wechel, filling most of a very large folio, some 50 cm. in height, which permitted greater clarity as well as more space for annotation and text. The two were more integrated than in earlier anatomical treatises by the placing of letters on the plates referring to the text, as well as enabling the easy inclusion of any new discoveries in both visual and verbal form. The last three images, showing the skeleton, were drawn by Calcar and based on the skeleton of a young man that had been articulated and prepared for him by Vesalius himself.[169] Their general thrust was still Galenic and relied on animals far more than upon human dissections, but in both text and illustration, Vesalius modified or cast doubt on certain of Galen's anatomical and physiological conclusions.[170] Galen's description of the coccyx and sacrum he found inconsistent, while he strongly doubted his claim that the lower jaw was divided into two. The prostate gland is also depicted for the first time.[171] But a major aim of the plates was to give the names of parts of the body in Latin, Greek and, occasionally in Arabic and Hebrew, thus allowing them to be used by those brought up in the pre-humanist traditions of medicine, so that the reader could easily locate them in the body.[172] Their derivation from Vesalius' lectures is also clear from the several comments that are made in the text concerning bloodletting and the relevant veins, a topic on which he had differed strongly from Drivère and on which he was to expand further in later lectures in Bologna and Padua and in his 1539 *Letter* on bloodletting.[173] In short, these plates and the accompanying text provided a more detailed visualisation of the new Galenic anatomy than ever before, even if, within a matter of months, Vesalius himself was to doubt, and even reject, some of what he had written here.

Vesalius was also a pioneer in that he worked closely for at least most of the *Tabulae* with a celebrated artist. As we shall see later, he had an artist's eye, and his upbringing in the penumbra of the highly cultured court of Brussels will have strengthened his aesthetic sensibilities as well as introducing him to many of the leading artists of the region. He and his family seem to have had links with the Stella family of artists, and it is possible that the artist he chose to work with in 1538 was already known to him.[174] Jan Stephan van Calcar (ca. 1515–ca. 1546) was a native of Kalkar, a small town only a few kilometres from Wesel, where Vesalius' family had originated, and where semi-distant members of the family were for long prominent.[175] Around 1532, he moved to Venice, where he joined the circle of Titian, but he was no impecunious artist, if he was able to pay for the production of the *Tabulae* as is stated on the final image.[176] This is his earliest datable work and its success will have enhanced both his career and that of Vesalius. There were other teachers of anatomy who engaged distinguished artists. Da Vinci's association with Marcantonio Della Torre has already been noted, and the Ferrarese anatomist Giambattista Canani (1515–1579) secured the services of Girolamo da Carpi (ca. 1501–1556) to illustrate his own anatomy book, but only the section on the muscles of the arms was completed by 1542.[177] Estienne may have been helped by some Italian artists from the school of Fontainebleau, but the extent of their contribution is unclear and may be at best only at second or third-hand.[178] But Vesalius' collaboration with Calcar was longer lasting and, in the end, more significant for both men, for Calcar went on to have a successful, albeit short, career in Naples and to be the subject of brief mentions in the second edition of Vasari's celebrated *Lives of the Painters*.[179]

Vesalius' innovation was quickly followed by other printers eager to cash in both on the images and on a market for similar productions for a wide readership. His plates were plagiarised by printers in Augsburg, Strasbourg, Frankfurt, Cologne, Paris and Marburg, to his great annoyance, and they appear to have stimulated the publication of so-called fugitive anatomical sheets.[180] These were pairs of single sheets, one male and one female, containing a relatively brief text as well as, sometimes, images of individual organs which could then be stuck down in order as flaps (hence the alternative name of flap anatomies) that could then be raised to show what was beneath them and give an idea of three-dimensionality (Fig. 4.15). That their utility was universally and swiftly acknowledged is clear from Vesalius' unusual decision in 1543 to insert an over-large folio showing the veins and arteries into the *Fabrica*, accompanied by a second folio depicting organs along with detailed instructions for cutting them out and sticking them down in order.[181] Both were taken over from the *Epitome*, the second in particular because it was prepared a little later than the *Fabrica* and was thus a little more accurate.[182] Not every copy preserves this or the later insert on the nerves on p. 353/453, and none of their owners followed Vesalius' advice to create a flap anatomy.[183] Some flap anatomies, like those published at Wittenberg between 1573 and 1625, were intended for use by students attending a course of lectures, but many were printed in places like Augsburg and London which had no university and were intended for wider consumption.[184] They continued to appear until the early years of the next century. Their relationship to Vesalius' originals is often remote, but in some of the Wittenberg prints, where they accompanied lectures based on Melanchthon's, *De anima*, *On the Soul*, it is jokingly indicated by the resemblance of the head of the male figure to that of Vesalius himself as portrayed in the *Fabrica* (Fig. 4.15).[185]

Although in his first anatomical demonstration, Vesalius had been content to refer his students to Guinter's book, their demand for something better, he claimed, encouraged him to bring out a revision of it in time for his next course.[186] His choice of printer fell on Bernardino Stagnino (1483–1538), a Venetian with many connections within the University of Padua.[187] The title page of the revision announced that it had been written by Guinter and enlarged and improved by Vesalius, an unexceptional formulation save that Vesalius' name, printed in large capitals, stood out and overshadowed that of his master.[188] Vesalius' appeal in his preface to Guinter's kindness shows that he was aware of the reaction that it might provoke from him and other Parisians.[189] Renaissance ideas about plagiarism and unauthorised publication are not ours, but his revision without Guinter's approval was unprecedented, not least because he was prepared to admit that the decision was his, not that of the publisher.

In his Preface, Vesalius offered two, not entirely convincing, justifications for publishing a revised edition. The first was the poor quality of the printing of the first edition, but Bernardino's was little better, as the long list of *Errata* demonstrates. Some errors in the first edition remained uncorrected, and new ones resulted from a failure to read Vesalius' handwriting or from a misunderstanding of what he had meant.[190] At p. 5v Bernardino had reused a matrix that he had just used for a page in a student edition of Galen's *De causis symptomatum*, *The Causes of Symptoms*, but while resetting the type at the top of the page, he had forgotten to remove the superfluous text at the bottom.[191] The mistake was not picked up by the author or printer until it was too late to change or even to mention it in the *Errata*.

Vesalius' second complaint about Guinter's little book was the haste with which it had been written. This is not unfair, for there are some obvious gaps, and the last book is somewhat disorganised and lacks the stylistic polish of the earlier ones. But it is in general a clear guide to dissection, well put together with an eye to what students might need. Vesalius, as we have already seen, was keen to get his message across to students in the best possible way, and his decision to produce an enlarged and corrected version of an effective textbook was understandable, particularly if he was eager to have something to give his students when the next academic year began.[192]

But the greatest impetus towards a revision came from his own discoveries that supplemented or contradicted those of his master. He emphasises his experience with the knife in three leading universities, f. + iii^{r}, and many of his corrections clearly derive from what he had done there, or in his private dissections of quadrupeds and reptiles.[193] To an even greater extent than Berengario and Massa he refers to what he himself could see or touch, adding his own empirical observations to the obvious book-learning of his master. They supplemented what Galen had written, but, as yet, he could accommodate his discoveries within a relatively flexible Galenic framework, as others also were doing, without the need to repudiate his great Greek predecessor. At the same time, the precision of his manner of dissection, as well as his remarkably acute eye, which have regularly drawn the admiration of modern anatomists familiar only with his work on the *Fabrica*, will have further enhanced his reputation as a technical genius.

Vesalius' corrections are extensive and go far beyond simple proof corrections or what earlier historians of medicine have identified. Some are relatively trivial, like his correction of the printer's misunderstanding of the letter S at f. 32v, or his rewording of the order of muscles in the introduction at f. 7r-v to correspond to the description that follows, but many others supplement Guinter from his own experience of dissecting both animals and man, or, as with his advice on suturing wounds, f. 9r, and on ascites, f. 13v, treating these conditions in practice. Often, as with the description of the reproductive system, ff. 29r–42v, small additions and corrections reveal a considerably greater experience acquired over a small number of years.[194] A few examples suffice to indicate their range.

There are new pieces of advice on how to carry out a dissection. The corpse should first be placed on its back, shaved of all hair, and then the umbilicus removed for use later in the demonstration f. 6v. It is not at all easy to separate the tendons of the second pair of muscles in the abdomen, f. 8r, and one should be warned about mistaking the different layers of the abdomen and perineum, f. 9r. The anatomy of the throat is complex, and any incision should be attempted only after practice on large animals.[195] Vesalius does not hesitate to correct or qualify Guinter's advice. At f. 6v-7r, he says that one need not bother with removing the membrane immediately under the skin, as Guinter appears to suggest, for this is essential only in dissecting a dog. Similarly, he dissents from Guinter's advice that the heart should always be removed from the thoracic cavity before proceeding with rest of the dissection, 58r. Sometimes, Vesalius suggests a better way of proceeding, as in evacuating water from between the peritoneum and the intestines, f. 13v, and he describes a new experiment to show how much the stomach can be distended by the intake of food, f. 57v. After cutting the trachea, one should insert a straw into the lung and have the assistant inflate it: even better, first one should cut the oesophagus and tie the end of the stomach at the duodenum. Dissecting the muscles of the shoulder girdle is relatively easy if you follow Vesalius' advice, beginning with the corpse on its back and then turning it over to complete the removal of the muscles.[196] In another record of his own experience, this time during the cutting open of the cranium, he recommends the insertion of a little stick or probe into the cut to tell whether one has reached the dura mater, for clumsy work will damage the delicate structures of the brain inside the skull f. 73r. In the same place he also criticises the Marburg professor Dryander's new method of opening the skull with a sickle-shaped tool instead of a thin straight saw.[197] His method of dealing with the eyeball, which is carefully explained, would appear to be his own invention, f. 84r, but not everyone would wish or be able to follow his sound advice, f. 96r, that the small nerves in the arm are best observed in rotting flesh.[198]

Vesalius also made new observations.[199] The ends of the straight abdominal muscles are sometimes crossed over so that the right one is attached to the left pubic bone, and vice versa, f. 8v, and there is a membrane at the ends of the ureters that allows fluid to drip but prevents any backflow, f. 33r. He indicates the position of the *rectus abdominis* muscle more precisely, f. 8v, and provides new information about the carotid arteries, f. 55v, the length of the vagina, f. 37r, and the muscles of the hands, f. 90v, and feet, f.102r.[200] A comment on the excision of the labia minora by Egyptians, f. 37r, shows his reading as well as an interest in the subject that can be traced well down into the 1550s.[201] His use of animals in dissection is more extensive and more sophisticated than in Guinter's original. The position of the omentum is often different in humans from in monkeys, f. 14r; the caecum in dormice has a remarkable capacity to expand, f. 15r; and the liver in reptiles takes up much of the left side, f. 21r. In humans, unlike in animals, the lung can adhere to the left side of the thorax, f. 61v.[202] At f. 53v. Vesalius' comment on the azygos vein in monkeys and dogs as well as in man not only expands on what Guinter had said, but also seems to lay claim to the discovery being originally his, possibly indicated to Guinter during a Parisian anatomy.[203]

The great majority of Vesalius' changes are additions rather than corrections, each no more than two or three lines long. But there are also larger sections added by Vesalius. A single paragraph devoted by Guinter to the branches of the portal veins is expanded over several pages into a description also of the arterial system of the mesentery, ff. 24v-27v, along with advice on how to dissect.[204] Much more information is now given on the muscles of the shoulder, ff. 46r-47r, and of the thigh, ff. 99r-v., and on those attached to the hyoid bone, ff. 66v-67r. Other passages raise questions that he will discuss further in his later work, such as the relationship between pulsation and the movements of the heart, f. 58v. He is also more aware of the difficulties in generalising from one or two dissections, telling his readers at f. 93r that they should not be too concerned about the different pattern of veins on the back of the hand.[205]

He was particularly proud of his discovery that there was a second membrane below the *pericranios*, the *periostios*, which was a covering common to all bones. (Fig. 2.7) At the beginning of Book III, p. 106, Guinter, discussing the skull, had declared that beneath the fleshy, muscular layer of the scalp there was a membrane called both *pericranios* and *periostios*, enclosing the skull on both sides and originating in the *dura mater.*[206] Once this layer had been scraped away, the skull itself could be clearly seen beneath. It was an opinion that went back to Galen and was subsequently accepted by all writers. In his revision Vesalius deleted the first mention of the *periostios* and began f. 72v with the observation that a second membrane could be found below the *pericranios*, the *periostios*, "which is a covering common to all bones": *Sub ea periost{r}<i>os omnibus ossibus corporis commune tegumentum est*, thus distinguishing the membranous *galea aponeurotica* from the true *pericranium.* Later, in his annotations, in a somewhat cramped note in the upper margin of f. 72v (Fig. 2.7), Vesalius added that up until now this had never been noticed by anyone: *hactenus non est ab aliquo animaduersa*, making even more explicit the claim that this underlying layer was his own discovery.[207] His demonstration of these layers at Bologna in 1540 was greeted enthusiastically by the Italians present, according to one of the participants.[208] Later, when he came to discuss the skull in the *Fabrica*, he repeated his belief that the pericranium and the periosteum were two different things, albeit in a somewhat more flowery form of words.[209]

That he may have felt a little embarrassed by his treatment of Guinter is further hinted at by his initial reluctance to respond to a request from one of his senior colleagues at Padua, Giambattista Da Monte (1498–1551), and from the editor of the Juntine press in Venice, Agostino Gadaldino (1515–75), to participate in their proposed edition of the collected works of Galen in Latin translation. This was a project that was in 'honourable rivalry', to use the words of his housemate at Padua, John Caius, with a similar one organised by another Paduan professor, Vittore Trincavella (1496–1568) and a younger doctor from Pavia, Agostino Ricci (1512–64). Vesalius' role was to revise the versions of some of Galen's anatomical works, including those by Guinter, in accordance with the best scholarly practice of the time. According to John Caius (1510–1573), Gadaldino provided them with an old Greek manuscript that they could check in

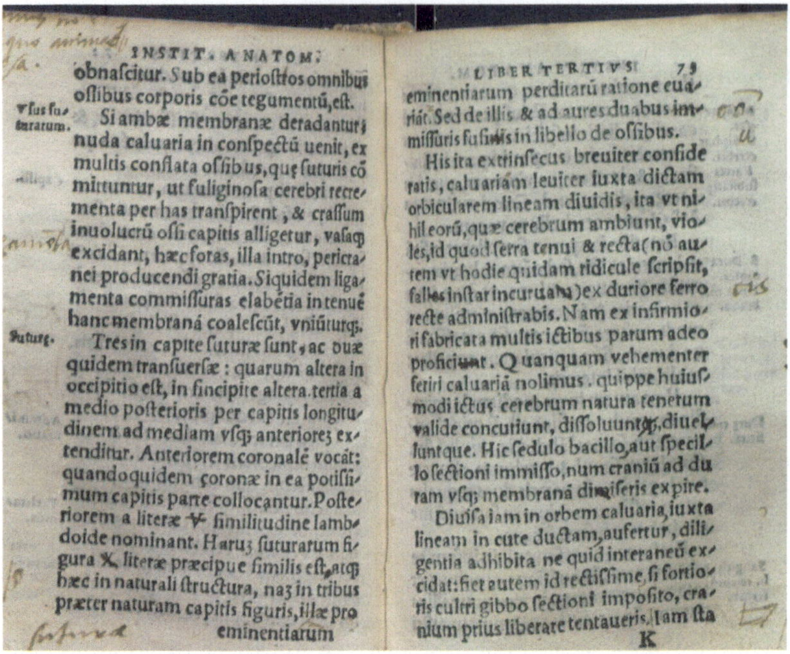
INSTIT. ANATOM.

obnascitur. Sub ea periosthos omnibus
ossibus corporis cōe tegumentū, est.
Vsus suturarum. Si ambæ membranæ detrahantur,
nuda caluaria in conspectū uenit, ex
multis conflata ossibus, quę suturis cō
mittuntur, ut fuliginosa cerebri recre-
menta per has transpirent, & crassum
inuolucrū ossi capitis alligetur, vasaq;
excidant, hæc foras, illa intro, pericra-
nei producendi gratia. Siquidem liga-
menta commissuras elabētia in tenuē
hanc membranā coalescūt, vniūturq;.
Suturę. Tres in capite suturæ sunt, ac duæ
quidem transuersæ: quarum altera in
occipitio est, in sincipite altera. tertia a
medio posterioris per capitis longitu-
dinem ad mediam vsq; anteriorē ex-
tenditur. Anteriorem coronalē vocāt:
quandoquidem coronæ in ea potissi-
mum capitis parte collocantur. Poste-
riorem a literæ Λ similitudine lamb-
doide nominant. Haruȝ suturarum fi-
gura X literæ præcipue similis est, atq;
hæc in naturali structura, naȝ in tribus
præter naturam capitis figuris, illæ pro
eminentiarum

LIBER TERTIVS 73

eminentiarum perditarū ratione eua-
riāt. Sed de illis & ad aures duabus im-
missuris fusius in libello de ossibus.
His ita extrinsecus breuiter conside
ratis, caluariam leuiter iuxta dictam
orbicularem lineam diuidis, ita vt ni-
hil eorū, quæ cerebrum ambiunt, vio-
les, id quod serra tenui & recta (nō au-
tem vt hodie quidam ridicule scripsit,
falles instar incuruata) ex duriore ferro
recte administrabis. Nam ex infirmio-
ri fabricata multis ictibus parum adeo
proficiunt. Quanquam vehementer
feriri caluariā nolimus. quippe huius
modi ictus cerebrum natura tenerum
valide concutiunt, dissoluuntq;, diuel-
luntque. Hic sedulo bacillo, aut specil-
lo sectioni immisso, num craniū ad du
ram vsq; membranā dimiseris expire.
Diuisa iam in orbem caluaria, iuxta
lineam in cute ductam, aufertur, dili-
gentia adhibita ne quid interaneū ex-
cidat: fiet autem id rectissime, si fortio-
ris cultri gibbo sectioni imposito, cra-
nium prius liberare tentaueris. Iam sta
K

Fig. 2.7 Andreas Vesalius and Johann Guinter, *Institutionum anatomicarum secundum Galeni sententiam... libri*, Venice: D. Bernardino, 1538. f. 72v. At the top left, Vesalius adds that the discovery of the *periostios* membrane beneath the skull was his own. (Copyright: Stuart Rose)

order to make the Latin more accurate.[210] But it was full of mistakes and when he and Vesalius studied it together, Caius concluded that it was worthless.[211] Vesalius made several improvements over what had been printed before, especially to Balamio's translation of *On Bones*, but these were mostly stylistic changes rather than modifications of Galen, and he does not seem to have consulted a larger manuscript of *Anatomical Procedures* that was then available in Venice, now Biblioteca Marciana, gr. 2379. But he was no philologist, and his version of *Anatomical Procedures* was roundly criticised later by Caius, who filled the margins of his copy of the 1538 Greek Galen with many imprecations against "Wesalius", which he amplified further in the notes and annotations to his own Greek edition in 1544. His much later reference in his *Autobibliography* to their co-operation was more conciliatory, cunningly suggesting that Vesalius eventually came to share many of his own attitudes towards Galen.[212] At least one other of Vesalius' readers was angered by what he saw as his incompetence as a philologist. The distinguished Galenist translator, Janus Cornarius, in his copy of the 1542 Basle edition of Galen's *Opera Omnia*, now in the British Library, classmark 9065–9 (2), crossed out Vesalius' name with such violence that the impression can still be seen several pages later. He was no less annoyed at what he later found in the *Fabrica*, preferring to emend all away the passages in Galen and Aristotle that Vesalius had singled out for criticism.[213]

Moving away from Galenic Anatomy

It is important to stress that at this date Vesalius was still proceeding within the parameters acceptable to most followers of Galen (and that in his therapeutics throughout his life he followed the Galenic tradition). Certainly, in his revision of the *Institutiones* that was published by Winter in Basle in 1539, Guinter accepted almost all the corrections that Vesalius had made to Galen, going beyond him in adding major sections on the anatomy of the head and neck and on the muscles of the arm and leg.[214] He included a passage on the muscles of the larynx, where Vesalius had explicitly criticised Galen, and removed another where Galen is praised for his discovery of the muscles in the nostrils.[215] But the great majority of the changes were either stylistic or removed what could be easily considered irrelevant. Others indicate disagreements over how to dissect a corpse or new information that Guinter had obtained with the assistance of his dissector, the later theological radical Miguel Servetus (1509?–1553). But throughout, Guinter displayed respect for his former helper, retaining his original praise of his work, describing him as "remarkably diligent in dissection, and no less capable in pure medicine". Indeed, his reprinting of the commentary has "performed an outstanding service".[216] The only suggestion of any displeasure may be found on the title page, where there is no mention of Vesalius and the book is described as "newly revised and augmented by the author himself".

The value of Vesalius' reworking of the *Institutiones* as a student guide can be seen in the decision of another Venetian publisher, Melchiorre Sessa, to reprint it at an unknown date.[217] A handwritten note by a near-contemporary on the title page in the Wellcome copy, EPB 2968, gives the date 1538, but this may well derive from a confusion with the original Basle printing. A more likely date is around 1540, by which time Vesalius was considerably occupied with the *Fabrica* and appears to have abandoned his plan to produce yet another edition.

The evidence for this change of tack is provided by the annotations that he made to his own copy, at present in a private library.[218] The great majority are simple proof corrections, but others engage in dialogue with a future printer. In one he asks for a reference to Galen to be placed in the margin "if there is space available", and in another, he asks for a comment to be placed in the margin with an asterisk to show that the error was Galen's, not his.[219] He adds words for clarification as well as suggesting marginal headings to help the reader. There are also a handful of tiny additions that may be the result of dissections and autopsies carried out since the first publication. At f. 37r,9, the covering of the female testes is now described as a slender membrane, while at f. 72v,9 Vesalius adds "ligaments" (*vasa <et ligamenta>*) to the vessels that pass through the skull to help form the pericranium. On the previous page, line 13, Guinter had briefly noted the different muscular structures of the forehead and cheeks and compared the latter with the lips. Vesalius added that there were also nerves there but cautioned that they were not all formed in the same way: in illa (namely fronte) *<enim> musculosa compago est, {in his} <ex> musculorum <et neruorum> fibris vt in labris, <non tamen eodem modo>*. To the slightly misleading description of the heart at f. 59v,8, Vesalius first added after "*vena*" the words "*ingreditur et venosa*", next deleted them, and finally rewrote the sentence more clearly at the bottom of the page: *caua vena ingreditur, vena vero arteriosa egreditur* ("The vena cava enters, and the artery-like vein leaves"). In the discussion of the leg at f. 96v, lines 21 and 23, Guinter's description of the tendon drawing the tibia upwards (*sursum trahens*) is altered to "turning it forwards" (*[in] anteriora vertens*).[220]

Vesalius had thought of writing his own anatomy book as early as 1538, for he apologises in the printed *Institutiones* for not saying more at this point about the muscles of the neck and spine, "reserving them for another place, since they cannot be dealt with in a brief account".[221] How large the book would be would depend on the number of bodies that he was able to dissect, but it would certainly be illustrated. He had begun work on the illustrations later in the year. At the end of the *Bloodletting Letter*, composed in late 1538, he revealed to his former patron, Nicholas Florenas, that he had already completed two illustrations of the cranial nerves and the small branches of the dorsal spine, but was unwilling to proceed further until he had completed drawings of the muscles and of all the internal parts. He also expressed the hope that Calcar would join him in his project and that they would be able to work together.[222] John Caius later also recalled that during their long stay together in 1539–40. Vesalius was writing and making the illustrations for the *Fabrica* (although one need not imagine that the text, and certainly his plan for the images, had as yet taken on the form they were to have in 1543).[223]

He may have been still hoping to produce a new edition of the *Institutiones*, but two considerations suggest that by late 1539 or 1540, Vesalius had abandoned all interest in it. Where one can identify later entries among his marginal annotations, either by changes in the style of script or position on the page, they resemble more aide-mémoires to passages that would later be discussed at length in the *Fabrica* than an update of what was already there (Fig. 2.8). He several times included specific references to Galen by page and section (the standard way of citing Galen rather than by line on a large folio), but they are to translations from the 1530s, not to anything in the Juntine edition of 1541–2.[224] Secondly, the latest datable reference in his notes would appear to be a comment squeezed into the top of fol. 72v, in which he emphasised that the discovery of the double covering of the skull was his alone, and "never before discovered by anyone else" (Fig. 2.7), almost certainly a riposte to Guinter's repetition in 1539 of Vesalius' original wording of his discovery.[225]

Furthermore, the more opportunities he had to dissect human bodies, the more discrepancies he found with what Galen had said. He later claimed that it was only after giving three courses of lectures on Galen's *On Bones* that he felt able to challenge his authority. Further proof of his move away from Galen is provided by a student's summary of his anatomical demonstrations in January 1540, when he was invited to Bologna to act as the dissector for the anatomy lectures of Matteo Corti (1475–1544). Corti had met Vesalius in Bologna once before, in 1538, when they had argued over the existence of fibres in the veins, which Vesalius believed he could show by dissection and thereby prove Galen right. Corti, by contrast, despite being a devoted adherent of Galen, thought Galen had here erred.[226]

Two years later the positions were reversed. Vesalius had been invited by the university students at Bologna to dissect three human bodies, six dogs and other animals to accompany Corti's lecture course on

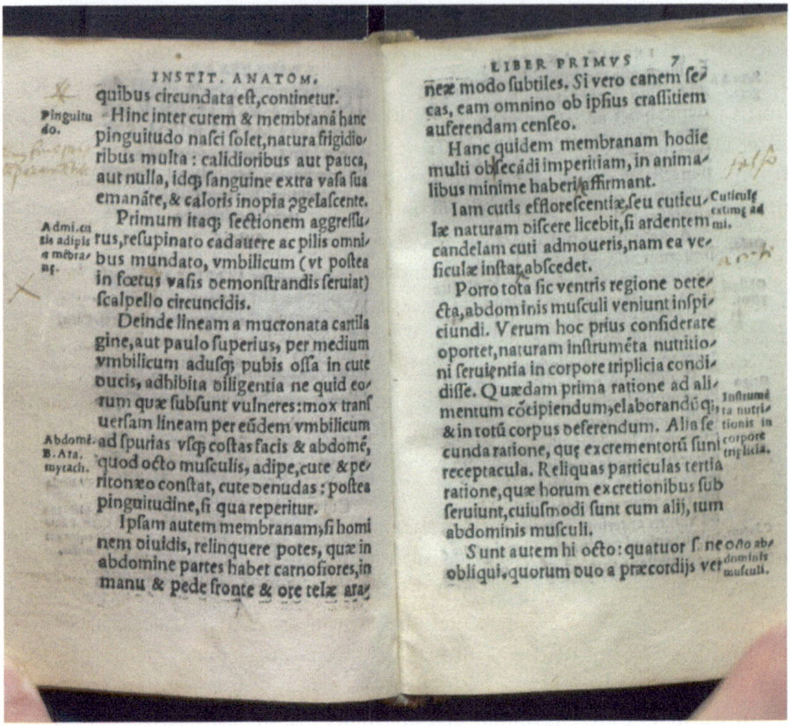
INSTIT. ANATOM.

quibus circundata eſt, continetur.
Pinguitudo. Hinc inter cutem & membranã hanc
pinguitudo naſci ſolet, natura frigidio-
ribus multa: calidioribus aut pauca,
aut nulla, idq; ſanguine extra vaſa ſua
emanãte, & caloris inopia ꝯgelaſcente.
Admi. cutis adipis & mẽbranæ. Primum itaq; ſectionem aggreſſu-
rus, reſupinato cadauere ac pilis omni-
bus mundato, vmbilicum (vt poſtea
in fœtus vaſis demonſtrandis ſeruiat)
ſcalpello circuncidis.
Deinde lineam a mucronata cartila
gine, aut paulo ſuperius, per medium
vmbilicum aduſq; pubis oſſa in cute
ducis, adhibita diligentia ne quid eo-
rum quæ ſubſunt vulneres: mox tranſ
uerſam lineam per eũdem vmbilicum
Abdomẽ. B. Ara. mytach. ad ſpurias vſq; coſtas facis & abdomẽ,
quod octo muſculis, adipe, cute & pe-
ritonæo conſtat, cute denudas: poſtea
pinguitudine, ſi qua reperitur.
Ipſam autem membranam, ſi homi
nem diuidis, relinquere potes, quæ in
abdomine partes habet carnoſiores, in
manu & pede fronte & ore telæ ara-

LIBER PRIMVS 7

neæ modo ſubtiles. Si vero canem ſe-
cas, eam omnino ob ipſius craſſitiem
auferendam cenſeo.
Hanc quidem membranam hodie
multi obſecãdi imperitiam, in anima-
libus minime haberi affirmant.
Cuticulæ extimæ admi. Iam cutis effloreſcentiæ, ſeu cuticu-
læ naturam diſcere licebit, ſi ardentem
candelam cuti admoueris, nam ea ve-
ſiculæ inſtar abſcedet.
Porro tota ſic ventris regione dete-
cta, abdominis muſculi veniunt inſpi-
ciundi. Verum hoc prius conſiderare
oportet, naturam inſtrumẽta nutritio-
ni ſeruientia in corpore triplicia condi-
diſſe. Quædam prima ratione ad ali-
Inſtrumẽta nutritionis in corpore triplicia. mentum cõcipiendum, elaborandũq;,
& in totũ corpus deferendum. Alia ſe
cunda ratione, quę excrementorũ ſunt
receptacula. Reliquas particulas tertia
ratione, quæ horum excretionibus ſub
ſeruiunt, cuiuſmodi ſunt cum alij, tum
abdominis muſculi.
Octo abdominis muſculi. Sunt autem hi octo: quatuor ſ ne
obliqui, quorum duo a præcordijs ver

Fig. 2.8 Andreas Vesalius and Johann Guinter, *Institutionum anatomicarum secundum Galeni sententiam… libri*, Venice: D. Bernardino, 1538, ff.6v-7r. Vesalius stresses others' mistakes (*falso*) and adds page references to an edition of Galen. (Copyright: Stuart Rose)

Mondino's *Anatomy*. He was to deploy his technical expertise in dissection while the senior man discoursed on the wider implications of the anatomy, correcting Mondino's medieval notions by reference to the newly rediscovered anatomical works of Galen. Conscious of his status, Corti suggested that Vesalius was a mere dissector, a little superior to a surgeon but scarcely a respectable physician if he could handle a scalpel so expertly. Their confrontation before an audience of almost 200 crowded into the church of San Francesco was certainly lively. The two men argued and interrupted one another. Vesalius often let Galenic doctrine pass without comment, but other points, like Galen's belief in the five lobes of the liver (which he had himself depicted in the *Tabulae*) or the precise point of insertion of the *rectus abdominis* muscle, he denounced as contrary to the actual evidence of dissection. He also used drawings and plates to confirm the significance of what could be seen, showing that it often contradicted what Galen had written. Such an open assertion of his own independence so annoyed Corti that he walked out in disgust at his temerity.[227] Vesalius stayed on until the end of the month to perform several private dissections, of both animals and humans, and mounted for Professor Bianchi skeletons of an ape and a French priest in order to compare their articulation.[228]

The significance of the Bologna debate is considerable. In it, Vesalius expressed strongly in public, possibly for the first time outside Padua, his conviction that mere book-learning was not enough and that demonstrable evidence should take precedence over what had been previously stated. It marked a stage on his move away from a narrow Galenism, but only a stage, and it did not, as yet, distinguish him unduly from those, like Sylvius, Corti, Berengario da Carpi or Niccolò Massa, who were prepared to believe that Galen had made this or that mistake but was still in general correct. Nor were they bothered by Galen's confession that he was forced to rely largely on animals for his description of the body, for they all, Vesalius included, inspected animals to supplement the supply of human corpses and bones. A general adherence to Galenic theory by Vesalius is clear from his letter on venesection of 1539, in which he supported Galen's rules on bloodletting against those of his Arabic interpreters, and from the Galenic therapeutics in his case notes from this period.[229]

Meanwhile, Vesalius' career in Padua flourished. His salary was increased in 1542 to 200 florins, and he was joined as a lecturer in surgery by his pupil Realdo Colombo. Yet increasingly he was occupied with his big book, the *Fabrica*, which he had begun in 1538 and which was substantially completed by summer 1542.[230] It was based first on his knowledge of Galenic anatomy and physiology and secondly on the evidence he had gleaned from his many dissections. Some bodies he dissected in public, like the mistress of a Paduan monk stolen from her tomb by eager students (possibly the female corpse depicted on the frontispiece of the *Fabrica*); others he had autopsied, like the oarsman of a papal trireme; still others, with the active encouragement of the local authorities, he had taken home to dissect in private after a public execution.[231] When human corpses and bones were lacking, he turned to animals to show the workings of the living body and where Galen had gone wrong. He spoke to midwives and to professors, he visited the butcher's stalls and the hospitals, he read or re-read his Galen, sometimes in Greek manuscripts, and he discussed his plans with fellow scholars and with artists. Above all, he wrote, and wrote, free from the distractions of a busy practice and taking advantage of the intellectual climate of Padua and nearby Venice.[232] He was a young man in a hurry, with a career to make.

Notes

1. Details of his family are given by C. Donald O'Malley, *Andreas Vesalius of Brussels (1514–1564)* (Berkeley and London: University of California Press, 1964), 21–27.
2. The alternative name was strongly advised by Guido Sold on the basis of documents in the ecclesiastical archives of Xanten in a poster session at the 2023 Antwerp Vesalius conference. I am grateful to him for further discussion on that occasion.
3. Stresa, Centro Internazionale di studi Rosminiani, classmark SIC a2 21, fol. 1r, reproduced in facsimile, Brussels: Academia Regia Belgica Medicinae, Dissertationes Series Historica 6, 1998; O'Malley, *Andreas Vesalius*, 23–25.
4. Andreas Vesalius, *Epistola, rationem modumque propinandi radicis Chynae decocti….pertractans* (Basle, J. Oporinus, 1546), 196–97. Sold, above, note 2, emphasised his lifelong links with his Rhineland relatives, who inherited his property there.
5. O'Malley, *Andreas Vesalius*, 25, doubting Vesalius' own story that he died aged 36.
6. Vesalius, *Epistola Chynae*, 196–97; for his use of his grandfather's manuscripts in the *Paraphrasis*, above, p. 24.
7. Vivian Nutton, *Renaissance Medicine. A Short History of Medicine in the Sixteenth Century* (London and New York: Routledge, 2022), 139–43.
8. Girolamo Cardano, *Libelli quinque* (Nuremberg: J. Petreius, 1547), 178r. For the possible alternative date, the next day, see O'Malley, *Andreas Vesalius*, 423, n. 2. Philip McNair, "Poliziano's Horoscope," in *Cultural Aspects of the Italian Renaissance,* edited by Cecil H. Clough (Manchester: Manchester University Press, 1976), 270, suggests that the horoscope was not cast by Cardano but by Joachim Rheticus (1514–1576).

9. O'Malley, *Andreas Vesalius*, 29–32, who also emphasises his training in Latin before going to Louvain; Andrea Carlino, "Medical Humanism: Rhetoric and Anatomy at Padua, circa 1540," in *Rhetoric and Medicine in Early Modern Europe*, edited by Stephen Pender and Nancy S. Struever (Farnham and Burlington, VT: Ashgate, 2012), 111–28.
10. Jean De Tournes, the publisher of the 1552 Lyon edition of the *Fabrica*, sig. a 1v, signalled out its language for its clarity and its propriety.
11. Andreas Vesalius, *Paraphrasis in nonum librum Rhazae medici Arabis clariss. ad Regem Almansorem, de singularum corporis partium affectuum curatione* (Basle: R. Winter, 1537?), 144: *Epistola Chynae*, 197; trans. Daniel H. Garrison, *Vesalius:* The China Root Epistle. *A New Translation and Critical* Edition (Cambridge: Cambridge University Press, 2115), 230.
12. Andreas Vesalius, *Paraphrasis in nonum librum Rhazae medici Arabis clariss. ad Regem Almansorem, de singularum corporis partium affectuum* curatione (Louvain: R. Rescius, 1537), sig. A iir v.
13. Elly Cockx-Indesteghe, "Jaspar Laet Senior, Govaert Bac en Pierre le Caron. Een pronosticatie voor het Jaar 1500," in *Hellinga, Festschrift, Festbundel, Mélanges*: *Forty-three Studies in Bibliography presented to Prof. Dr. Wytze Hellinga*, edited by Arnold R.A. Croiset van Uchelen (Amsterdam: Nico Israel, 1980), 74–80.
14. Hubert Barland, *Epistola ad medicinae apud Lovanienese studiosam juventutem*, in Giovanni Manardo, *Medicinales epistolae* (Strasbourg: J. Schott. 1529), sig. a.iir: Henry De Vocht, *History of the Foundation and the Rise of the Collegium Trilingue Lovaniense* 1517–1550) (Louvain: Bibliotheque de l'Université, 1951–55), II, 516–24.
15. Some of his comments in the margin of the revised *Paraphrasis* to herbs in shops may refer to his experiences in Paris or in Flanders, *ParaphrasisBasle* 38, 105, 110, 112, 221–22.
16. Laurence Brockliss and Colin Jones, *The Medical World of Early Modern* France (Oxford: Clarendon Press, 1997), 85–95, 131–33, 170–79.
17. Nutton, *Renaissance Medicine*, 99–113.

18. Michael D. Reeve, "Celsus", in *Texts and Transmission. A Survey of the Latin Classics*, edited by Leighton D. Reynolds (Oxford: Clarendon Press, 1983), 46–47. For Celsus as a medical author, Vivian Nutton, *Ancient Medicine*, ed. 3 (London and New York: Routledge, 2024), 131–32.
19. Nigel G. Wilson, *From Byzantium to Italy. Greek Studies in the Italian Renaissance* (London: Duckworth, 1992).
20. Stefania Fortuna, "L'insegnamento di Nicolò Leoniceno e la sua incidenza nella storia della medicina italiana ed europea," in *Nicolò Leoniceno 1428 – 1524. Un umanista veneto nella storia della medicina. Atti del Convegno in Lonigo per il 590 anniversario della nascita*, edited by Alberto Lonigo (Contro: Riccardo editore, 2019), 81–96; Raphael Ebgi, "Cum barbaris decertare. *Sull'umanesimo medico di Niccolò Leoniceno e sulla sua controversia col Poliziano*," *Schifanoia* 50/51 (2016), 55–62.
21. Daniela Mugnai Carrara, "Profilo di Nicolò Leoniceno," *Interpres* 2 (1979), 188; Alberto Lonigo (ed.), *Nicolò Leoniceno 1428 – 1524. Un umanista veneto nella storia della medicina. Atti del Convegno in Lonigo per il 590 anniversario della nascita* (Contro: Riccardo editore, 2019).
22. Vivian Nutton, "The Rise of Medical Humanism: Ferrara, 1464–1555," *Renaissance Studies* 11 (1997), 2–19.
23. For medics and the humanism of the court, see Douglas Biow, Doctors, *Ambassadors, and Secretaries: Humanism and Professions in Renaissance Italy* (Chicago and London: The University of Chicago Press), 2002; Maria Kavvadia, "Sources and Resources of Court Medicine in Mid-Sixteenth Century Rome: Erudition as an Epistemological and Ethical Claim," in *Scientiae in the History of Medicine*, edited by Fabrizio Baldassari and Fabio Zampieri (Rome and Bristol, CT: L'Erma di Bretschneider, 2021), 171–88, concentrates on the importance of erudition for court doctors such as Vesalius.
24. Daniela Mugnai Carrara, *La biblioteca di Nicolò Leoniceno tra Aristotele e Galeno: cultura e libri di un medico umanista* (Florence: L. S. Olschki, 1991), 35: of the items inventoried in his library, in both print and manuscript, at his death, 247 out of 340 concerned medicine, philosophy or mathematics.

25. Niccolò Leoniceno, *De Plinii et aliorum in medicina* erroribus (Ferrara: L. de Rubeis and A. de Grassis, 1492), sigg, a.3v, 4r, 4v. for Arabs and modern druggists, sigg. C .3v–4r.
26. Francis Maddison, Margaret Pelling and Charles Webster (eds), *Essays on the Life and Work of Thomas Linacre* c.1460–1524 (Oxford: Clarendon Press, 1977).
27. This and the following paragraph summarise Brockliss and Jones. *Early Modern France*, 94–120.
28. Emil Arbenz, ed. *Die Vadianische Briefsammlung der Stadtbibliothek St. Gallen, II, Mitteilungen zur vaterländische Geschichte* 25, no. 2 (1891), 238.
29. Marie-Louise Concasty, *Commentaires de la Faculté de médecine de l'Université de Paris (1516–1560)* (Paris: Imprimerie Nationale), 100.
30. Jean Fernel, *De naturali parte* medicina (Paris: A. Wechel, 1542) is his account of the structures and, to a much greater extent, the processes of the human body. See also the introduction to Jean Fernel, *The* Physiologia *of Jean Fernel (1567). Translated and Annotated by John M. Forrester with an Introduction by John Henry and John M. Forrester* (Philadelphia: American Philosophical Society, 2003).
31. Cyril E. Kellett, "Sylvius and the Reform of Anatomy," *Medical History* 5 (1961), 101–16; Gerhard Baader. "Jacques Dubois as a Practitioner," in *The Medical Renaissance of the Sixteenth Century*, edited by Andrew Wear, Roger K. French and Iain M. Lonie (Cambridge: Cambridge University Press, 1985), 146–54. A proper study of Sylvius and his writings is a desideratum.
32. Jacobus Sylvius, *Methodus sex librorum Galeni in differentiis et causis morborum et symptomatum in tabellas sex ordine sua coniecta... de signis omnibus* medicis...commentarius (Paris: C. Wechel, 1539); *Ordo et ordinis ratio in legendis Hipporatis et Galeni libris* (Paris: C. Wechel, 1539).
33. Jacobus Sylvius, *In Hippocratis et Galeni physiologiae partem anatomicam* Isagoge (Paris: J. Hulpeau, 1555). Galen's description of the limbs is more accurate than that of the internal organs.
34. Vivian Nutton and Christine Nutton, "Noël du Fail, Cardano, and the Paris Medical Faculty," *Medical History* 48 (2004), 367–72.

35. Vesalius, *Paraphrasis* 1537: f. 3 v.
36. In all he translated over forty tracts of Galen, many for the first time, as well as portions of the Hippocratic Corpus and Paul of Aegina. A brief biography in Vivian Nutton, *Johann Guinter and Andreas Vesalius, Principles of Anatomy According to the Opinion of Galen* (London and New York: Routledge, 2017), 22–27.
37. For a possible earlier dissection, Vivian Nutton and Christine Nutton, "The Archer of Meudon: a Curious Absence of Continuity in the History of Medicine," *Journal of the History of Medicine and Allied Sciences* 58 (2003), 401–27.
38. The history of anatomical dissection from Antiquity onwards is substantial of which only a selection can be given here. In general, Kenneth B. Roberts and John D. W. Tomlinson, *The Fabric of the Body. European Traditions of Anatomical Illustration* (Oxford: Oxford University Press, 1992); Mimi Cazort, Monique Kornell, and Kenneth B. Roberts, *The Ingenious Machine of Nature* (Ottawa: National Gallery of Canada, 1996); Monique N. Kornell, *Flesh and Bones: The Art of Anatomy* (Los Angeles: Getty Trust Publications, 2022); for Antiquity, Julius Rocca, "Anatomy," in *The Cambridge Companion to Galen*, edited by R. James Hankinson (Cambridge: Cambridge University Press, 2008), 242–62; Claire Bubb, *Dissection in Classical Antiquity: A Social and Medical* History (Cambridge: Cambridge University Press, 2022); for the Later Middle Ages, Katharine Park. "The Criminal and the Saintly Body: Autopsy and Dissection in Renaissance Italy," *Renaissance Quarterly* 47 (1994), 1–33; *Secrets of Women: Gender, Generation and the Origins of Human Dissection* (London: Zone Books, 2006); for the Renaissance, Roger K. French, *Dissection and Vivisection in the European Renaissance* (Aldershot and Brookfield VT: Ashgate, 1999); Andrea Carlino, *Books of the Body: Anatomical Ritual and Renaissance Learning* (Chicago: Chicago University Press, 1999); Giuseppe Olmi and Claudia Pancino, eds. A*natome: sezione, scomposizione, raffigurazione del corpo nell'età moderna* (Bologna: Bononia University Press, 2012); Nutton, *Renaissance Medicine*, 245–77, and for the wider cultural background, Jonathan Sawday, *The Body Emblazoned. Dissection and the Body in Renaissance* Culture (London: Routledge, 1995); David Hillman and Carla Mazzio, eds., *The Body in Parts: Fantasies of Corporeality in Early Modern*

Europe (London and New York: Routledge); Richard Sugg, "The Anatomical Web: Literary Dissection from Castiglione to Cromwell," in *Rhetoric and Medicine in Early Modern Europe*, edited by Stephen Pender and Nancy S. Struever (Farnham and Burlington, VT: Ashgate, 2012), 83–103.

39. Mondino de' Liuzzi. *Anothomia*, edited by Petro P. Giorgi and Giorgio F. Pasini (Bologna: Istituto per la Storia dell' Università di Bologna, 1992); Taylor McCall, *The Art of Anatomy in Medieval* Europe (London: Reaktion Books, 2023), 98–101. Mondino stresses his considerable experience of dissecting humans, presumably in autopsies outside the context of university lectures. See also Kira Robison, *Healers in the Making: Students, Physicians, and Medical Eduction in Medieval Bologna (1Det-1550)* (Leiden and Boston: Brill, 2021), 121–23.
40. Nutton, *Renaissance Medicine*, 247–53.
41. For Galenic anatomy, see Julius Rocca, *Galen on the Brain. Anatomical Knowledge and Physiological Speculation in the Second Century AD* (Leiden: Brill, 2003); Luis A. Salas, *Cutting Words: Polemical Dimensions of Galen's Anatomical Experiment* (Leiden and Boston: Brill, 2020); Vivian Nutton, *Galen: A Thinking Doctor in Imperial* Rome (London and New York: Routledge, 2020); "Galen and the Latin *De Voce*: a new Edition and English Translation," in *Nell' officina del filologo. Studi sul testo e i loro lettori per Ivan Garofalo*, edited by Tommaso Raioli and Amneris Roselli (Pisa and Rome: Fabrizio Serra, 2022), 141–64: Bubb, *Dissection*, 274–348.
42. Théodore Morel, *Tredecim orationes encomiasticae habitae Parisiis, Anno M.D.*XXIII (Paris: N. Savetier, 1523); Vivian Nutton, "André Vésale et l'anatomie parisienne," *Cahiers de l'Association Internationale des Études Françaises* 55 (2003), 239–49.
43. Noel Du Fail, *Contes* facétieuses (Paris: P. Daffis, 1874), 145–46, probably referring to lectures in 1539.
44. Galen, *De anatomicis* administrationibus (Paris: S. De Colines, 1531); *Libri anatomici* (Bologna: G. B. Phaelli, 1529).
45. Vesalius, *Episotla Chynae*, 152; tr. Garrison, *China Root*, 181. The version in O'Malley, *Andreas Vesalius*, 51, is defective. At *Epistola Chynae*, 141: tr. Garrison, *China Root*, 169, he talks of attending a formal dissection inexpertly carried out on two

corpses and ending within three days, which must have taken place in 1534.

46. For the private dissections, Nutton, *Johann Guinter*, 168 (a marginal addition by Vesalius); Johan Guinter, *De medicina veteri et nova facienda* (Basle: ex officina Henricpetrina, 1571), 261, mentions vivisecting dogs. Vesalius was succeeded as dissector for Guinter by the Spaniard Miguel Servetus, famous for being burnt at the stake for his heretical *Restitutio Christianismi*.
47. Andreas Vesalius, *De humani corporis fabrica* (Basle: J. Oporintus, 1543), 538.
48. Ibid., 44, 159; *Epistola Chynae*, 194; tr. Garrison, *China Root*, 227.
49. Vesalius, *Fabrica*, 159.
50. Ibid., f. *3r.
51. Jean Tagault, *De chirurgica* institutione (Paris: Chrestien Wechel, 1543).
52. Andreas Laguna, *Anatomica methodus, seu de sectione humani corporis contemplatio* (Paris: J. Kerver, 1536), 58–59.
53. Charles Estienne, *De dissectione partium corporis* humani (Paris: S. de Colines, 1545). A French version reusing the same plates was published in 1546.
54. Bette Talvacchia, *Taking Positions. On the Erotic in Renaissance Culture* (Princeton: Princeton University Press, 1999), concentrates on the relationship of Estienne's images to what preceded them in art and literature, but says little about the medical context, for which see Roberts and Tomlinson, *The Fabric of the Body*, 166–87. Hélène Cazes and Andrea Carlino, "Plaisir de l'anatomie, plaisir du livre: *La dissection des parties du corps humain* par Charles Estienne (Paris, 1546), *Cahiers de l'Association Internationale des Études Françaises* 55 (2003), 251–69, discuss the wider French cultural context.
55. Pierre Huard and Mrko Grmek, *L'oeuvre de Charles Estienne et l'école anatomique* parisienne (Paris: Au cercle du livre précieux, 1965), 4, report, via Renouard, an *Anatomia* in 8° mentioned in a catalogue of 1536 by Robert de Colines, of which no copy has yet been found. But, unless this is a phantom and the result of some bibliographer's error, its size suggests that it had little to do with what appeared in 1545 but would have been a short and unillustrated handbook like those of Guinter and Laguna. Roberts

and Tomlinson, *Fabric of the Body*, 176, suggest a probable chronology.

56. Talvacchia, *Taking Positions*, 182–44; Huard and Grmek, *Estienne*, 5, suppose that the four early plates, Estienne, *De dissectione*, 150, 151,154, 155, were drawn for a different publication before Estienne became interested in medicine.
57. For the importance of the illustrations, see below, 92–93.
58. O'Malley, *Andreas Vesalius*, 89–90. Roberts and Tomlinson, *Fabric of the Body*, 173, are sceptical of the extent and date of Estienne's borrowing from Vesalius.
59. Andreas Vesalius, *De humani corporis fabrica* (Basle: J. Oporinus), 161.
60. Hubert Barland, *Velitatio cum Arnoldo Nootz ...Deque Avicennae in plaerisque horum hallucinatione. Deinde obiter agitur de venae in pleuritide sectione. Postremo Avicennae locis aliquot aliis dormitatio ostenditur, simul Galenus ab hominis morsibus aliquot defenditur* (Antwerp: Henricus Petri of Middelburg, 1532).
61. Johann Guinter and Andreas Vesalius, *Institutionum anatomicarum secundum Galeni sententiam... libri ...Ab Andrea Wesalio Bruxellensi auctiores et emendatiores redditi* (Venice: D. Bernardino, 1538), sig. iii verso; Johann Guinter, *Principles of Anatomy According to the Opinion of Galen*, translated by Vivian Nutton (London and New York: Routledge, 2019), 62; Henry De Vocht, *History of the Foundation and the Rise of the Collegium Trilingue Lovaniense* 1517–1550 (Louvain: Bibliothèque de l'Université, 1951–55), III, 532–34.
62. Jérôme Drivère, *De missione sanguinis in pleuritide...commentarius de victu ab arthriticis morbis vindicante, ubi quam male diris illis cruciatibus sit a neotericis hactenus provisum,* ostenditur (Louvain: Rutger Rescius, 1532), t.p.
63. Andreas Vesalius, *Paraphrasis in nonum librum Rhazae medici Arabis clariss. ad Regem Almansorem, de singularum corporis partium affectuum curatione* (Basle: R. Winter, 1537?), 36, and the marginalia at 104, 180; *Epistola, docens venam axillarem dextri cubiti in dolore laterali secandam* (Basle: R. Winter, 1539).
64. Vesalius dedicated his 1538 revision of the *Principles of Anatomy* to him, Guinther and Vesalius, *Institutiones*, sigg. I–iii; tr. Nutton, *Johann Guinther*, 61–62.
65. *Fabrica*, 1543, fol. *3r.

66. For this and another dissection of a girl in noble service, Vesalius, *Epistola, rationem modumque propinandi radicis Chynae decocti….pertractans* (Basle: J. Oporinus, 1546), 141–42: translated by Daniel H. Garrison, *Andreas Vesalius:* The China Root Epistle. *A New Translation and Critical Edition by Daniel H.* Garrison (Cambridge: Cambridge University Press, 2015), 169; Vesalius, *De humani corporis fabrica* (Basle: J. Oporinus), 1555: 658.
67. This story may have been the basis for the illustrated initial letter L in the *Fabrica*.
68. Nancy G. Siraisi, *Medieval and Early Modern Medicine. An Introduction to Knowledge and Practice* (Chicago: University of Chicago Press, 1990), 130–31.
69. Vesalius, *Epistola Chynae*, 195; tr. Garrison, *China Root*, 228–29.
70. That was why his revision and that of Alban Thorer of Book X took the place of Gerard's version in the 1544 Basle reprinting of the collected works of Rhazes. Frank Hieronymus, *Theophrast und Galen—Celsus und Paracelsus. Medizin, Naturphilosophie und Kirchenreform im Basler Buchdruck bis zum dreissigjährigen Krieg* (Basle: Universitätsbibliothek, 2005), 292, draws attention to Thorer's similarly paraphrastic style in some of his earlier translations. For Thorer (1489–1549), humanist and lecturer in Basle, see later, p. 103.
71. Marginal headings were common in medieval medical manuscripts, but indexes were extremely rare until the age of print.
72. Vesalius, *Paraphrasis*, 23r, 73r.
73. Ibid. 22r, 84r, 49r, 97r.
74. Ibid. 13r; 13r-v; 77r.
75. Ibid. 7r, 24r, 81r.
76. Ibid.: e.g. Galen 29v, 35r; Paul 9r, 57v; Aetius 9r; Alexander 12r; Celsus 35r; Pliny 25r; Scribonius 45v; Theophrastus 66r.
77. Vesalius, *Paraphrasis Basle*, 17r, 48v, 71v, 72v 48v. Moritz Roth, *Andreas Vesalius Bruxellensis* (Berlin: Georg Reimer, 1892), 76, n. 3, argued for his use of the 1526 edition of Manardo's *Epistulae medicinales*.
78. Vesalius, *Paraphrasis Basle*, pharmacies, ibid. 18r, 54v, 67r, 106r; 'many people' 100v; 'we' 34r, 47r, 58r, 61r, 72v, 76v.
79. Ibid. 34r, 47r, 52r, 58r, 61r, 72v, 76v.

80. Ibid., 7v, 103r. To whom Vesalius is referring as 'modern' (*recentiores*) is far from clear. Post-Galenic authors such as Paul, Aetius and Alexander are usually named in his comments, but it is also difficult to imagine him meeting Greek surgeons in Paris: had he been given this information by others who had spent time in Venice (with its large Greek community) or further East?
81. Maurits Biesbrouck, "Rutgerus Giffeus, un nouvel ami d'André Vésale, qui fait de la Paraphrasis de la BIUM un exemplaire unique au monde," *Histoire des Sciences médicales* 45, no. 2 (2011), 196–99.
82. An observation by Maurits Biesbruck, to whom this section owes much. The word 'Theologiae' in the heading also lacks the first 'o', a further sign that this was very much a hurried job. Presumably, the BNF copy was that given to Giffeus.
83. Erasmus, *Ep.* 867, but, excluding overnight stays, around a fortnight; Percy S. Allen, ed., *Opus epistolarum Des. Erasmi* Roterodami (Oxford: Clarendon Press, 1906–1947), III 393–401.
84. O'Malley, *Andreas Vesalius*, 72.
85. Harvey M. Cushing, *A Bio-Bibliography of Andreas Vesalius* (Hamden, CT, and London: Archon Books, 1962), 6, although in the end he rejected it. Abdul Haq Compier, "Rhazes in the Renaissance of Andreas Vesalius," *Medical History* 56 (2012), 13, is alone in accepting this later date.
86. For Platter's editions of Galen in March 1537 and March 1538, see Hieronymus, *Theophrast und Galen*, 84–5, 95–6; nos 47 and 50.
87. Roth, *Vesalius*, 76, note 1.
88. Maurits Biesbrouck and Omer Steeno, "The *Paraphrasis ad librum nonum Avicennae*, first book by Vesalius," in *La Fabrique de Vésale et autres textes*, on line at https://www.biusante.parisdescartes.fr/vesale/, consulted 14/09/24.
89. O'Malley, *Andreas Vesalius*, 72.
90. Below, pp. 172, 180.
91. This number does not include the expansions of references in the first edition to earlier medical texts.
92. Vesalius, *ParaphrasisBasle*, respectively 168, 53, 111, 116, 180, 84, 96; 116; 91.

93. Ibid., respectively 131, 141, 145, 177; 9, 132, 173. Arculanus' commentary was printed in 1493 and 1497 alongside the text of Rhazes.
94. Ibid., respectively 9, 189; 144 (referring to the dung of wild (?) doves). At p. 83, he compares readings in various 'exemplars' (the word can refer equally to manuscripts and printed books).
95. Ibid., 32, 109, 180.
96. Ibid., 38, 105, 110, 112, 221–222 (*officinae*); 71, 73, 90 (occasionally useful); 140, 167 (superseded); to be avoided, 149, 153, 158, 213.
97. Ibid., 86–87, 135, 140; Brabant, 160. Cf. 189, confirming the late date.
98. Ibid., 'We' or I', 43, 52. 65, 77, 82, 86–87, 143, 161, 167–68, 197, 201 and 216.
99. Cf. Michael A. Screech,"Histoire des idées et histoire du livre: une optique personnelle et une profession de foi," in *Le Livre dans l'Europe de la Renaissance. Actes du xxviiie Colloque international d'Etudes humanistes de Tours*, edited by Pierre Aquilon and Henri-Jean Martin, (Nantes: PROMODIS, 1988), 554–66. For an overall view of medical printing, see Andrew Pettegree, *The Book in the Renaissance* (New Haven: Yale University Press, 2010), 297–318.
100. Jacopo Berengario da Carpi, *Isagoge breves...in anatomiam humani corporis* (Bologna: H. de Benedictis, 1522; *Isagoge breves...in anatomiam humani corporis* (Bologna: H. de Benedictis, 1523).
101. Janus Cornarius, *Universae rei medicae ἐπιγραφή seu enumeratio, compendio tractata* (Basle: Froben, 1529); *Universae rei medicae ἐπιγραφή seu enumeratio, compendio tractata, eius auctore recognita atque aliquot locis aucta* (Basle: Froben, 1534).
102. Leonhart Fuchs, *Compendiaria ac succincta admodum in artem medendi....introductio* (Hagenau: J. Secerius, 1531); *Compendiaria in artem medendi introductio, diligenter recognita, & locis multis quam antea auctio* (Strasbourg: J. Albertus, 1534).
103. He may well have planned to take his revised text with him to Padua in the hope of finding a printer there or in Venice.
104. Frank Hieronymus, "Physicians and Publishers in Sixteenth-century Basle," in *The German Book 1450–750. Studies Presented to David L. Paisey on his Retirement*, edited by John L. Flood and

William A. Kelly (London: The British Library, 1995), 95–110; Hieronymus, *Theophrast und Galen*, especially the chronological list, pp. 3695–96.

105. Christoph Reske, *Die Buchdrucker des 16. Und 17. Jahrhunderts im deutschen Sprachgebiet. Auf der Grundlage des gleichnamigen Werke von Josef Benzing, 2. überarbeitete und erweiterte Auflage* (Wiesbaden: Harrassowitz, 2015), 78–84, gives brief biographies of the four men. Martin Steinmann, *Johannes Oporinus, ein Basler Buchdrucker um die Mitte des 16. Jahrhundert* (Basle and Stuttgart: Helbing & Lichtenhahn, 1967), 10–12, documents problems within the group.
106. The Basle Index of Basle Publishers, on-line at https://ub.unibas.ch/de/historische-bestaende/alte-drucke-438; consulted 15/9.22. queries the 1536 date for Winter's printing of Pomponius Mela, cf. Reske, *Buchdrucker*, 78. The 1537 printing of a work by the French humanist Dolet was done by Platter on behalf of Winter alone, a further indication that he was striking out on his own as a publisher.
107. Nutton, *Johann* Guinther, 22–27; Hieronymus, *Theophrast und Galen*, 84–85.
108. Steinmann, *Oporinus*, 108; Alfred Hartmann, Beat Jenny, Reinhard Bodenmann and Lorenz Heiligensetzer, *Die Amerbachkorrespondenz*, 11 vols in 14 (Basle: Universitätsbibliothek, 1942–2010), vol. 8, (1974), xxiv–xxv.
109. One might surmise that Vesalius' original intention was not to publish in Basle but in the far more prestigious Venice once he had reached Italy, which would have allowed time for yet more annotations.
110. This avoids the objection to the book being sent from Padua at a busy time of Vesalius' life and appearing in March 1538 (which remains a possibility, if somewhat unlikely).
111. Cynthia Klestinec, "Medical Education in Padua: Students, Faculty and Facilities," in *Centres of Medical Excellence: Medical Travel and Education in Europe, 1500–1789*, edited by Ole P. Grell, Andrew Cunningham, and Jon Arrizabalaga (Farnham: Ashgate, 2010), 193–211.
112. Giuseppe Ongaro, "La medicina nello Studio di Padova e nel Veneto," in *Storia della cultura veneta. Vol. III.: Dal primo Quattrocento al concilio di Trento*, edited by Gianfranco Folena

(Vicenza: Neri Pozza Editore, Part 3, 1981), 118–26; Paul F. Grendler, *The Universities in the Italian* Renaissance (London and Baltimore: Johns Hopkins University Press, 2002), 32–33.

113. For Bologna, Marinella Pigozzi, "Bologna. Dall'anatomia agli esemplari del corpo," in *Anatome: sezione, scomposizione, raffigurazione del corpo nell'età moderna*, edited by Giuseppe Olmi and Claudia Pancino (Bologna: Bononia University Press, 2012), 87–116; David Lines, *The Dynamics of Learning in Early Modern* Italy (Cambridge, MA, and London: Harvard University Press, 2023, 236–45, 250–57.
114. Harvey M. Cushing, *A Bio-Bibliography of Andreas Vesalius* (Hamden, CT, and London: Archon Books, 1962), 6. Falloppia, for instance, taught at Ferrara, Pisa and Padua for several years before taking his medical doctorate in 1552, Michael Stolberg, *Gabrielle Falloppia 1522/3–1562. The Life and Work of a Renaissance* Anatomist (London and New York: Routledge, 2023), 22–25.
115. Andreas Vesalius, *De humani corporis fabrica* (Basle: J. Oporinus, 1543), 538. 2023: 25.
116. Camillo Semenzato, *The Anatomy Theatre. History and* Restoration (Padua: Università degli Studi, 1995); Paula Findlen, "Anatomy Theatres, Botanic Gardens, and Natural History Collections," in *The Cambridge History of Science, vol. 3, Early Modern Science*, edited by Katharine Park and Lorraine Daston (Cambridge: Cambridge University Press, 2006), 274–78; cf. also Vesalius, *Fabrica*, 548.
117. The older, alternative, view, is that the professor from his dais gave the lecture while someone else pointed to the relevant part at appropriate points in the demonstration.
118. For *De juvamentis*, Roger K. French, "*De Juvamentis Membrorum* and the Reception of Galenic Physiological Anatomy," *Isis* 70 (1979), 96–109; *De motibus,* as Mondino knew it, was a Latin version, via Arabic, of a genuine work, Vivian Nutton with Gerrit Bos, *Galen, On Problematical* Movements (Cambridge: Cambridge University Press, 2011)*; De voce,* a collection of extracts from a lost Galen work, Vivian Nutton, "Galen and the Latin *De Voce*: a new Edition and English Translation," in *Nell' officina del filologo. Studi sul testo e i loro lettori per*

Ivan Garofalo, edited by Tommaso Raioli and Amneris Roselli (Pisa and Rome: Fabrizio Serra, 2022), 141–64.

119. Anna Maria Ieraci Bio, "Giorgio Valla e la medicina. Disegni anatomici e nomenclatura delle parti del corpo umano nel Mutin. α. W.5.5," *Galenos* 11 (2017), 227–45.
120. Giorgio Valla, *De expetendis et fugiendis rebus* (Venice: Aldus Manutius, 1501); Paola Landucci Ruffo,"Le fonti della medicina nell'enciclopedia di Giorgio Valla," in *Giorgio Valla tra scienza e sapienza*, edited by Vittore Branca (Florence: Olschki, 1981), 55–68; Amadeo A. Raschieri, "Giorgio Valla, Editor and Translator of Ancient Scientific Texts," in *Greek Science in the Long Run. Essays on the Greek Scientific Tradition (4th c. BCE-16th c. CE)*, edited by Paula Olmos (Newcastle: Cambridge Scholars Publishing, 2112), 127–49; Ieraci-Bio, "Giorgio Valla". For Vesalius' use of it, *Fabrica*, 547.
121. Alessandro Benedetti, *Historia corporis sive Anatomice* (Venice: B. Guerraldo, 1502); Giovanna Ferrari, *L'esperienza del passato. Alessandro Benedetti, filologo e medico umanista* (Florence: L. S. Olschki, 1996); (ed.), *Alessandro Benedetti*, Historia corporis sive Anatomice (Florence: Giunti, 1998); Carlo Giacomelli, "Medicina Patavina. Codici di medicina a Padova, fra Bessarione, Niccolò Leonico Tomeo e Marco Antonio della Torre (?)", *Revue d'histoire des textes* 16 (2012), 75–113.
122. Gabriele Zerbi, *Liber anathomiae corporis humani et singulorum membrorum illius* (Venice: B. Locatello, 1502); Roger K. French, *Dissection and Vivisection in the European Renaissance* (Aldershot and Brookfield VT: Ashgate, 1999), 81–91. Although both anatomical books were printed in the same year, both men had long been familiar with each other's arguments much earlier. For Zerbi's work on medical ethics, see Richard Tait, "The Editions of Gabriele Zerbi's *De cautelis medicorum* and their Influence", *Journal of the Warburg and Courtauld Institutes* 83 (2020), 327–36.
123. Antonio Benivieni, *De abditis nonnullis et mirandis morborum et sanationum causis* (Florence: F. Giunta, 1507). For the importance of autopsies in developing anatomical knowledge, see Katharine Park, "The Criminal and the Saintly Body: Autopsy and Dissection in Renaissance Italy," *Renaissance Quarterly* 47 (1994), 1–33; Victoria von Hoffmann, "Epistemologies of Touch

in Early Modern Autopsies," *Renaissance Quarterly* 75 (2022), 542–582.

124. Jacopo Berengario da Carpi, *Commentaria....super Anatomia Mundini* (Bologna: H. de Benedictis, 1521); *Isagoge breves...in anatomiam humani corporis* (Bologna: H. de Benedictis, 1522); *A Short Introduction to Anatomy: Isagoge Breves. Translated by L. R. Lind with anatomical notes by P.G. Roofe* (Chicago: University of Chicago Press, 1959). See also Roger French, "Berengario da Carpi and the Use of Commentary in Anatomical Teaching," in *The Medical Renaissance of the Sixteenth Century*, edited by Andrew Wear, Roger French and Ian M. Lonie (Cambridge: Cambridge University Press, 1985), 42–74; *Dissection*, 91–101; Lines, *Dynamics*, 254–56. For the illustrations, below, p. 41.
125. Galen, *Libri anatomici* (Bologna: G. B. Phaelli, 1529).
126. His teacher, Alessandro Achillini (d. 1512), claimed to have seen the rete in 1503, but also raised questions about what exactly it was: Alessandro Achillini, *De humani corporis anatomia* (Venice: G. A. Di Sabio, 1521); Levi R. Lind, *Studies in Pre-Vesalian Anatomy; Biography, Translations, Documents* (Philadelphia: American Philosophical Society, 1975), 61; Douglas J. Lanska, "Evolution of the Myth of the Human Rete Mirabile Traced through Text and Illustrations in Printed Books: The Case of Vesalius and his Plagiarists," *Journal of the History of the Neurosciences* 31 (2022), 221–61.
127. Berengario, *Commentaria*. Most of the illustrations cluster in the middle of the book (four images of a woman) ff. CCXXVv–VIr, and eight of a male at the end, ff. CCCCXVIIIv–XXIIr. See also below, pp. 41–42.
128. Berengario, *Introductio*, 15–16.
129. Berengario, *Isagoge breves...in anatomiam humani corporis a communi medicorum academia usitatam, a Carpo in almo Bononiensi Gymnasio ordinariam chirurgiae docente, ad suorum scolasticorum preces in lucem datae* (Bologna: H. de Benedictis, 1523), t.p. The same claim was repeated on the title page of the 1535 edition, Venice: B. Vitale.
130. Vivian Nutton, "*Prisci dissectionum professores*: Greek Texts and Renaissance Anatomists." In *The Uses of Greek and Latin, Historical Essays*, edited by A. Carlotta Dionisotti, Anthony Grafton and Jill Kraye (London: The Warburg Institute, 1988), 111–26.

131. Lind, *Pre-Vesalian Anatomy*, remains the best selection of Renaissance anatomical writing in English translation. For artists, see below, p. 40. For Vesalius' colleague, Landi, see Giuseppe Ongaro, "Il *'De humana historia'* (1542) di Bassiano Landi," *Atti della VI Biennale della Marca per la storia della medicina (Fermo, 29 aprile-2 maggio, 1965)* (Fermo: n.p., 1965, 265–78; "Bassiano Landi e Andrea Vesalio," *Atti e memorie dell'Accademia patavina di scienze, lettere ed arti* 110, part 2 (1997–8), 33–54; Fabrizio Bigotti, "Commenting on Aristotle with a Knife. The Heretical Anatomies of Bassiano Landi," *NTM, n.s.* 31 (2023), 1–26.
132. Ruben Eriksson, *Andreas Vesalius' First Public Anatomy at Bologna 1540. An Eye-witness Report by Baldasar Heseler medicinae scolaris together with his Notes on Matthaeus Curtius' Lectures on Anatomia Mundini* (Uppsala and Stockholm: Almqvist and Wiksell, 1959); Antonio De Ferrari, "Corti, Matteo," in *Dizionario biografico degli italiani* (1960-) (Rome: Treccani, vol. 29, 1983), 795–97; Vivian Nutton, ""Qui magni Galeni doctrinam in re medica primus revocavit" Matteo Corti und der Galenismus im medizinischen Unterricht der Renaissance," in *Der Humanismus und die oberen Fakultäten, Mitteilungen XIV der Kommission für Humanismusforschung*, edited by Gundolf Keil, Bernd Moeller and Winfried Trusen (Weinheim: VDA, 1987), 173–84.
133. Niccolò Massa, *Epistolae medicinales et philosophicae* (Venice: F. Bindoni and M. Pasini, 1550), *Ep.* 1,6: f. 58r (written in 1544 as a review of the *Fabrica)*. He may have had a point, even if somewhat exaggerated. See also Richard J. Palmer, "Nicolò Massa, his Family and his Fortune," *Medical History* 25 (1981), 385–410.
134. Francis Van Glabbeek and Maurits Biesbrouck. "Giovanni Baptista Canani (1515–1579) and Andreas Vesalius (1514–1564)," in *In the Shadow of Vesalius*, edited by Rob Van Hee (Antwerp and Apeldoorn: Garant, 2020), 101–48: Francis Van Glabbeek et al., *Myology before Vesalius. Giovanni Battista Canani (1515–1579 n.s.) and his* Musculorum Humani Corporis Picturata Dissectio (Antwerp and Apeldoorn: Garant), 28–29; Marco Bresadola, "*Giovanni Battista Canani: l'umanesimo medico e la riforma dell'anatomia rinascimentale.*" *Schifanoia* 50/51 (2016), 13–20.

135. François Rabelais, *Hippocratis et Galeni libri aliquot ex recognitione Francisci Rabelaesi* (Lyon: S. Gryphius, 1538); Estienne Dolet, *Carminum libri quatuor* (Lyon: [the author], 1538), *Carm.* IV, 18: p. 164; Roland Antonioli, *Rabelais et la médecine* (Geneva: Droz, 1976), 60–61,78–81, 288–89; Marie-Madeleine Fontaine, *Libertés et savoirs du corps à la Renaissance* (Caen: Paradigme, 1993), 195–220.
136. Vivian Nutton, "The Anatomy of the Soul in Early Renaissance Medicine," in *The Human Embryo. Aristotle and the Arabic and European Traditions*, edited by Gordon R. Dunstan (Exeter: University of Exeter Press, 1990), 136–57; "Wittenberg Anatomy," in. *Medicine and the Reformation*, edited by Ole P. Grell and Andrew Cunningham (London and New York: Routledge, 1993), 11–32; Jürgen Helm, "Religion and Medicine: Anatomical Education at Wittenberg and Ingoldstadt," in *Confessions and the Sciences in the Sixteenth Century*, edited by Jürgen Helm and Annette Winkelmann (Leiden: Brill. 2001), 51–68; David Cellamare, "History of a Productive Failure. The Use of Andreas Vesalius's *De humani corporis fabrica* in some Sixteenth-century Books on the Soul," in *Towards the Authority of Vesalius: Studies on Medicine and the Human Body from Antiquity to the Renaissance and Beyond*, edited by Erika Gielen and Michèle Goyens (Turnhout: Brepols, 2018), 403–36. Lutherans also strongly believed in the physical resurrection of the body.
137. Philip Melanchthon, *Briefwechsel. Kritische und kommentierte Gesamtausgabe* (Stuttgart and Bad Cannstatt: Verlag Frommann-Holzboog, 1977–), *Epp.*1182, 1430. In his earlier *Physics*, he had taken his medical information from Benedetti and Berengario.
138. O'Malley, *Andreas Vesalius*, 81–83 and pl. 16.
139. Andreas Vesalius, *Tabulae anatomicae sex* (Venice: B. Vitale, 1538a), fol. 1: R. Allen Shotwell, "Animals, Pictures and Skeletons. Andreas Vesalius's Reinvention of the Public Anatomy Lesson," *Journal of the History of Medicine and Allied Sciences* 71 (2016), 1–18; "The Illustrated Dissection: Drawings in the Extant Notes from Vesalius' Dissections," in *Towards the Authority of Vesalius: Studies on Medicine and the Human Body from Antiquity to the Renaissance and Beyond*, edited by Erika Gielen and Michèle Goyen (Turnhout: Brepols, 2018), 383–402.
140. Vesalius, *Tabulae*, fol. 1.

141. English translation, Charles Singer and Chaim Rabin, *A P*relude *to* Modern Science*: being a Discussion of the History, Sources and Circumstances of the 'Tabulae anatomicae sex' of* Vesalius (Oxford, Oxford University Press, 1946).
142. For contemporary debates about the truth value of visual representations, Martin Kemp, "The Mark of Truth: Looking and Learning in some Anatomical Illustrations from the Renaissance and the Enlightenment," in *Medicine and the Five Senses*, edited by William F. Bynum and Roy S. Porter (Cambridge: Cambridge University Pres, 1993), 85–121; Sachiko Kusukawa, *Picturing the Book of Nature. Image, Text and Argument in Sixteenth-Century Human Anatomy and Medical Biology* (Chicago: Chicago University Press, 2012), 162–77.
143. Vesalius, *Tabulae*, fol. 1.
144. Andrea *Carlino, Paper Bodies. A Catalogue of Anatomical Fugitive Sheets 1538–1687, Medical History Supplement* 19 (London: The Wellcome Institute, 1999), 46–58.
145. Loren C. MacKinney, "The Beginnings of Western Scientific Anatomy: New Evidence and a Revision in Interpretation of Mondeville's Role," *Medical History* 6 (1962), 233–39; McCall, *Art of* Anatomy, 83–86, 101–15, 123–24.
146. The style of the existing images in this manuscript, Florence, Biblioteca Medicea- Laurenziana, plut. 74,7, indicates a Byzantine reworking of an older and probably fourth- or fifth-century CE original. Apollonius' sketches will have been much less luxurious and perhaps fewer in number, Hermann Schöne, *Apollonius von Kitium* (Leipzig: B. G. Teubner, 1896); Massimo Bernabò, "La Rinascenza Macedone e l'arte dei bendaggi," *Kos* I, no. 5 (1984), 45–47.
147. Carlino, *Paper Bodies*, 74–81; McCall, *Art of Anatomy*, emphasises the different contexts of these illustrations and, 33–81, suggests a possible origin for the five-picture series within Benedictine monasteries.
148. Ibid., 81–88; Domenico P. Meli, "Images & Color: The Strasbourg Printer Johann Schott (1477–1548) and His Circle," *Early Science and Medicine* 27, no. 4 (2022), 539–40; McCall, *Art of Anatomy*, 150–54.

149. For Schott as an innovator, Miriam Usher Chrisman, *Lay Culture, Learned Culture. Books and Social Change in Strasbourg, 1480–1599* (New Haven and London: Yale University Press, 1982), 135–36; Meli, "Images".
150. Chrisman, *Lay Culture*, 135; Meli, "Images", 538–42 for his illustrations of surgery.
151. The famous Isenheim altarpiece by Matthias Grünewald was being painted in 1512–16 for a hospital near Colmar in Alsace and was sculpted by Nicholas of Haguenau, a town just north of Strasbourg.
152. Andreas Carlino, *Books of the Body: Anatomical Ritual and Renaissance Learning* (Chicago: Chicago University Press, 1999), 6–68, and McCall, *Art of Anatomy*, 163–75, discuss pre-Vesalian anatomy scenes.
153. For the development of book illustration in general, John Harthan, *The History of the Illustrated Book. The Western* Tradition (London: Thames and Hudson, 1980), 79–103; Jean-Paul Pittion, *Le livre à la Renaissance. Introduction à la bibliographie historique et matérielle* (Turnhout: Brepols, 2013), 183–210.
154. Monique N. Kornell, *Flesh and Bones: The Art of Anatomy* (Los Angeles: Getty Trust Publications, 2022), 62–63: for Salmon, and tentatively, Paulette Choné, "Le corps pathétique de Gabriel Salmon (v. 1485–1542)," in *La Représentation du corps à la Renaissance*, péristyles 42, edited by Paul Vert, Catherine Charoy and Mireille Mazet (Nancy: Musée des Beaux Arts, 2013), 105–16.
155. Bernard Schultz, *Art and Anatomy in Renaissance Italy* (Ann Arbor, Mich.: UMI Research Press, 1985); Monique N. Kornell, *Artists and the Study of Anatomy in Sixteenth-Century Italy* (PhD Diss. University of London, 1993); Domenico Laurenza, "Art and Anatomy in Renaissance Italy. Images from a Revolution," *Metropolitan Museum of Art Bulletin* 69, no. 3 (2012), 1–48, gives a good selection of images in his discussion of the interactions between artists and anatomists.
156. Domenico Laurenza, "In Search of a Phantom; Marcantonio della Torre and Leonardo's late Anatomical Studies," in *Leonardo's Anatomical World. Language, Context and "Disegno"*, edited by Alessandro Nova and Domenico Laurenza (Venice: Marsilio, 2011), 61–77. In general, Charles Nicoll, *Leonardo da Vinci. The*

Flight of the Mind (London: Penguin, 2004); Martin Clayton and Ron Philo, *Leonardo da Vinci:* Anatomist (London: Royal Collection Enterprises Ltd, 2012). Kenneth Keele and Carlo Pedretti, *The Drawings of Leonardo da Vinci in the Collection of Her Majesty the Queen at Windsor Castle* (New York: Harcourt Jovanovitch, 1978–80), remains fundamental.

157. This is not to say that earlier treatises including descriptions of the body were never illustrated, with varying degrees of realism, but they were few and often retained, like the famous manuscript of Guido de Vigevano of 1345, in a non-medical context, McCall, *Art of Anatomy*, 107–15.
158. Domenico Laurenza,"Contenuto, canoni visive e techniche nella illustrazioni anatomiche del Rinascimento. Riflessioni in margine ad una tavola di Berengario da Carpi." in *Anatome: sezione, scomposizione, raffigurazione del corpo nell'età moderna*, edited by Giuseppe Olmi and Claudia Pancino (Bologna: Bononia University Press, 2012), 27–44.
159. Ludwig Choulant, *History and Bibliography of Anatomic Illustration, translated and annotated by Mortimer Frank* (Chicago: University of Chicago, 1920). 138–42; Roberts and Tomlinson, *The Fabric of the Body*, 70–83; Carlino, *Books*, 18–22.
160. Vesalius, *Fabrica*, 114: Cushing, *Bio-Bibliography*, 10–12, with Figs. 10–13. O'Malley, *Andreas Vesalius*, 130, following Vesalius, *Fabrica*, 43. Wrongly gives the name of the Cardinal as Rodolfo and identifies him with Cardinal Rodolfo Pio of Carpi. The name may have been misremembered or misprinted, for the Cardinal was the Florentine Cardinal Ridolfi, see the edition of *De ossibus* by Ivan Garofalo and Armelle Debru, *Galien, Tome VII. Les os pour les débutants, L'anatomie des* muscles (Paris: Les Belles Lettres, 2005), 13.
161. Lucien Febvre and Henri-Jean Martin. *L'apparition du* livre (Paris: Albin Michel, 1999), 133–50; see more recently, Pittion, *Le livre*, 195–206; Sven Lembke, "Wie der menschliche Leichnam zu einem Buch der Natur ohne Druckfehler wird. Über den epistemologishen Wert anatomischer Sektionen im Zeitalter Vesal," in *Zergliederung—Anatomie und Wahrnehmunge in der Frühen Neuzeit*, edited by Albert Schirrmeister and Matthias Pozsgai (Frankfurt am Main: Vittorio Klostermann, 2005), 19–49; Sachiko Kusukawa, "*Ad vivum* Images and the Knowledge of

Nature in Early Modern Europe," in *Ad vivum? Visual Materials and the Vocabulary of Life-Likeness in Europe before 1800*, edited by Thomas Balfe, Joanna Woodall and Claus Zittel (Leiden and Boston: Brill, 2019), 89–121. I have not seen Judith Folkenberg, *L'anatomie humaine: cinq siècles de sciences et d'art* (Paris, Éditions de la Martinière, 2006).

162. Minta Collins, *Medieval Herbals. The Illustrative* Tradition (London: The British Library and The University of Toronto Press), 239–98.
163. Febvre and Martin, *L'apparition*, 386–87; Meli "Images".
164. Leonhart Fuchs, *De historia stirpium commentarii insignes* (Basle: M. Isingrin, 1542). For Isingrin's career, Reske, *Buchdrucker*, 73, who gives his first independent printing as 1532.
165. Frederick G. Meyer, Emily E. Trueblood, and John L Heller (eds). *The Great Herbal of Leonhart Fuchs: De historia stirpium commentarii insignes, 1542* (Stanford: Stanford University Press and Cambridge: Cambridge University Press, 1999), vol. 1; Kusukawa, *The Book of Nature*, 124–37; for Gessner and Mattioli, ibid., 162–77.
166. For possible influence, below, pp. 52–54.
167. Roth, *Vesalius*, 462: Carlino, *Paper Bodies*, 49–51.
168. Sten G. Lindberg, "Chrestien Wechel and Vesalius: Twelve Unique Medical Broadsides from the Sixteenth Century," *Lychnos* 1953, 50–74. The second edition may have incorporated some new material, but it agrees in general with what Choulant and Roth describe of the first.
169. For Vesalius' method of articulating a skeleton, see Monique N. Kornell, "Vesalius' Method of Articulating the Skeleton and a Drawing in the Collection of the Wellcome Library," *Medical History* 44 (2000), 97–110.
170. Singer and Rabin, *Prelude*, xliii-lxiv.
171. Ibid., Plate 1.
172. Ibid., lxv–lxxvi.
173. Ibid., Plate 1.
174. Franciscus Maria Stella, his cousin, acted for the family in the lawsuit against Oporinus' heirs, see below, p. 186.
175. Monique N. Kornell, "Jan Steven van Calcar (c. 1515–c. 1546), Vesalius's Illustrator," in *Andreas Vesalius and the "Fabrica" in the Age of Printing: Art, Anatomy, and Printing in the Italian*

Renaissance, edited by Rinaldo Canalis and Massimo Ciavolella (Turnhout: Brepols, 2018), 99–130.

176. This information is given on only one of the two surviving sets of plates. Some have suggested that he merely handed over Vesalius' money to the printer, others that both men contributed and that the printed sets that lacked Calcar's name were given to Vesalius. All assume that Calcar was an impoverished artist, which may not have been the case, even if his family were not as rich as that of Vesalius.
177. Francis Van Glabbeek, Maurits Biesbrouck and Jacqueline Vons, *Myology before Vesalius. Giovanni Battista Canani (1515–1579 n.s.) and his* Musculorum Humani Corporis Picturata Dissectio (Antwerp and Apeldoorn: Garant, 2022).
178. Monique N. Kornell, "Rosso Fiorentino and the Anatomical Text," The Burlington Magazine 131, no. 1041 (1989), 842–47.
179. Kornell, "Calcar".
180. Carlino, *Paper Bodies*, 46–73.
181. Jérôme Van Wijland, "La Fabrique du corps humain de l'œuvre," in Paul Vert, Catherine Charoy and Mireille Mazet, *La Représentation du corps à la Renaissance,* péristyles 42 (Nancy: Musée des Beaux Arts, 2013), 41–48.
182. Vesalius, *Fabrica,* 338/438.
183. Ibid., 313/413–16. The double number is necessary because the setter by error misnumbered the pages. For two surviving copies of the *Epitome* with the relevant pages turned into a flap anatomy, see Carlino, *Paper Bodies*, 61, n. 19. The London, Wellcome Collection copy of the 1555 edition, EPB 6562 (1), is bound with a pair of 1539 anatomical sheets, EPB 288, 3–4; Daniel Margócsy, Mark Somos and Stephen N Joffe. *The* Fabrica *of Andreas Vesalius. A Worldwide Descriptive Census, Ownership and Annotations of the 1543 and 1555 Edition* (Leiden and Boston: Brill, 2018), 402, but the binding is not original.
184. Carlino, *Paper Bodies*, 240–44; 120–32; 180–82.
185. Ibid., 240–4; 250–61.
186. Johann Guinter and Andreas Vesalius. *Institutionum anatomicarum secundum Galeni sententiam... libri ...Ab Andrea Wesalio Bruxellensi auctiores et emendatiores redditi* (Venice: D. Bernardino, 1538), sig. ii^{v}. It goes without saying that not all the changes introduced by Vesalius could be illustrated visually.

Vivian Nutton, *Johann Guinter and Andreas Vesalius, Principles of Anatomy According to the Opinion of* Galen (London and New York: Routledge, 2017), translates this edition, indicating the sections changed by Vesalius and providing in Appendix 1 a transcription of Vesalius' own annotations.

187. EDIT 16, online at https://edit16.iccu.sbn.it/ s.v. Stagnino, Bernardino, lists his printings in the years 1537–1538 of Brasavola's little book of syrups, Dioscorides, and five editions of works by Galen, four of Hippocrates and Mondino's *Anatomy* (most as small 16mo student editions), as well as this book.
188. A possible printing of this book by di Vitale, reported in Stuttgart and noted by Sachiko Kusukawa, *The Book of Nature*, 185, is the result of a cataloguer's misunderstanding of the name on the colophon.
189. Guinter and Vesalius, *Institutiones*, sig. ii$^{v.}$
190. For changes indicating an original marginal note by Vesalius misunderstood by the printer, see ibid., ff. 34v, 52v, 101r, and for a possible misreading of his handwriting, f. 99v. Others may be private comments taken over into the text.
191. Galen, *De morbis et symptomatis libri sex* (Venice, D. Bernardinus, 1538), sig. 5 verso. I am grateful to Stephen Greenberg of the National Library of Medicine, Bethesda, for checking the folio reference for me.
192. The speed with which he worked may also account for his failure to write up the promised section on the muscles of the neck and spine, Guinter and Vesalius, *Institutiones*, f. 85r. Cf. also f. 71r for Vesalius' summary attempt to fill in a gap deliberately left by Guinter when dealing with the muscles of the jaw.
193. Guinter and Vesalius, Institutiones, f. + iii^{r}.
194. And possibly also from reading, if Singer and Rabin, *Prelude*, lxi, are right to suspect a knowledge of Massa's description of the prostate gland in his *Anatomiae liber introductorius* (Venice: F. Bindoni and M. Pasini, 1536), ff. 34r–36r; tr. Lind, *Studies*, 198-201; Giorigo Weber, *Sensata Veritas* (Florence: Lero S. Olschki, 2006), reprints the book as an appendix to his study of Massa's and Colombo's pathological investigations. Massa later claimed, *Epistolae medicinales et philosophicae* (Venice: F. Bindoni and M. Pasini, 1550), *Ep.*, I,6: f. 58r, that the author of the *Fabrica* had

learned from him, and now thought himself more accomplished than the ancients.

195. Guinter and Vesalius, *Institutiones,* f. 66r. Cf. Galen, *De motibus dubiis*, 11, 21–2; edited by Vivian Nutton with Gerrit Bos, *Galen, On Problematical* Movements (Cambridge: Cambridge University Press, 2011), pp. 170–71.
196. An observation confirmed to me by Francis Van Glabbeek from his own teaching experience.
197. He will have known of this from Dryander's 1536 *Anatomia capitis humani* (Marburg: E. Cervicornus), sig. A iv[r].
198. Francis Van Glabbeek reminds me that in the 1540 Bologna demonstrations bodies had to be replaced when the stench became too much, and wonders whether Vesalius had some form of anosmia that allowed him to continue dissecting under these conditions.
199. Antoine Drizenko, "Les *Institutions Anatomiques* de Jean Guinter d'Andernach (1487–1564) et André Vésale (1524–1564)," *Histoire des sciences médicales* 45 (2011), 134, wrongly attributes the discovery of the pancreas of Aselli, f. 16r, to Vesalius: he had merely transferred a comment by Guinter, *Institutiones*, p. 17, to a different place. Guinter in his revision, p. 14, retained its original position.
200. Jennifer Kosmin, "Midwifery Anatomized: Vesalius. Dissection and Reproductive Authority in Early Modern Italy," *The Journal of Medieval and Early Modern Studies* 48 (2018), 79–104. But his strengthened acceptance of a two-chambered womb, f. 36r, also shows his ignorance of human female anatomy. He had revised his opinion by the time of the *Fabrica.*
201. Vivian Nutton, "Vesalius Revised: his Annotations to the 1555 *Fabrica*," *Medical History* 56 (2012), 429–31.
202. In the 1540 Bologna dissections Vesalius used dogs to show many parts of the body not available to him in a male corpse that had been already dissected.
203. Cf. Vesalius, *Tabulae*, Plate II, with Singer and Rabin, *Prelude*, lvi-lvii; Eriksson, *Heseler*, 272–73; Vesalius, *Fabrica*, 279–82/379–82. Guinter's original comment, *Institutiones*, 64–65-; tr. Nutton, *Johann Guinter,* 130, reads like an afterthought brought to his attention during a Vesalian dissection.

204. O'Malley, *Andreas Vesalius*, 94, wrongly extends the new section to include the description of the spleen on ff. 24r-v and describes the whole as solely referring to the spleen; he also wrongly claims, p. 94, that Vesalius had altered Guinter's discussion of the *rete mirabile*, f. 80r-v. Note also the substantial reworking, ff, 52v–54v, of Guinter's description of the vena cava.
205. Nancy G. Siraisi, *Medicine and the Italian Universities, 1250–1600* (Leiden: Brill, 2001), 287–327, shows how Vesalius developed this interest in variation in both editions of the *Fabrica.*
206. Guinter, Institutiones, 106.
207. The way in which this clause is written around an already existing addition suggests that it could have been written in response to Guinter's inclusion of his original comment unchanged in his own revision, *Institutionum anatomicarum secundum Galeni sententiam libri quatuor.* Basle: Robert Winter, 1539, 80.
208. Eriksson, *Heseler*, 218–19.
209. Vesalius, *Fabrica*, 627.
210. Ivan Garofalo, "Agostino Gadaldino (1515–1575) et le Galien latin," in *Lire les médecins grecs à la Renaissance: aux origines de l'édition médicale: actes du colloque international de Paris, 19–20 septembre 2003*, edited by Véronique Boudon-Millot and Guy Cobolet (Paris: Bibliothèque interuniversitaire de médecine, 2004), 283–322.
211. Galen, *Opera quae extant omnia* (Venice: Giunta, 1541–42), vol. 2, pp. 55v–59r, 63r–105v. For his use of manuscripts, see John Caius, *De libris propriis, liber* unus (London: W. Seres, 1570), 6r-v; trans. by Vivian Nutton, *An Autobibliography by John* Caius (London and New York: Routledge, 2018), 58.
212. Caius, ibid.
213. Andreas Vesalius, *Epistola, rationem modumque propinandi radicis Chynae decocti….pertractans* (Basle, J. Oporinus, 1546), 178: trans. Daniel H. Garrison, *Vesalius:* The China Root Epistle. *A New Translation and Critical* Edition (Cambridge: Cambridge University Press, 2015), 210. For Cornarius as a philologist, see Brigitte Mondrain," Éditer et traduire les médecins grecs au XVIe siècle: L'exemple de Janus Cornarius," in *Les voies de la science grecque: Études sur la transmission des textes de l'Antiquité au dix-neuvième siècle* (Paris: De Boccard, 1997), 391–417; Marie

Laure Montfort, *Janus Cornarius et la redécouverte d'Hippocrate à la Renaissance* (Turnhout: Brepols, 2018).

214. Guinter, Institutiones 1539, 96–102, 118–25.
215. Guinter, Institutiones 1539, 72 and 78 = Guinter and Vesalius, *Institutiones* 1538, ff. 65v and 70v.
216. Guinter, Institutiones 1539, sig. v^{r}. For more details of the changes, see Nutton, *Johann Guinter*, 42–45.
217. Johann Guinter and Andreas Vesalius. *Institutionum anatomicarum secundum Galeni sententiam... libri ...Ab Andrea Wesalio Bruxellensi auctiores et emendatiores redditi.* (Venice: M. Sessa, n.d.).
218. Details in Nutton, *Johann Guinter.* I am grateful to Stuart Rose for permission to examine his copy.
219. Ibid., 169, 172.
220. Ibid., 168–75.
221. Guinter and Vesalius, *Institutiones*, 85v.
222. Andreas Vesalius, *Epistola, docens venam axillarem dextri cubiti in dolore laterali secandam* (Basle: R. Winter, 1539), 64–65.
223. Caius, *De libris*, fol. 6v: tr. Nutton, *John Caius*, 58. His Paduan colleague Landi knew more of the images and grasped their significance by late 1541, Landi, *De humana historia*: 119.
224. Nutton, *John Caius*, 41.
225. Caius, *De libris*, fol. 6v: tr. Nutton, *John Caius*, 58.
226. Vesalius, *Epistola docens venam axillarem dextri cubiti in dolore laterali secandam* (Basle: R. Winter, 1539), 7, 19; *Anatomicarum Gabrielis Falloppii observationum examen* (Venice: Francesco de' Franceschi, 1564), 81–82.
227. Eriksson, *Heseler*, 272–73, 284–85.
228. Dates in Maurits Biesbrouck, Theodoor Goddeeris and Omer Steeno, "Andreas Vesalius in Bologna," in *Tijdgenooten uit de leefwereld van Andreas Vesalius*, edited by Rob Van Hee and Maurits Biesbrouck (Antwerp and Apeldoorn: Garant, 2018), 123–40.
229. Vesalius, *Epistola docens venam, passim;* for his earlier view, above, p. 19.
230. Ibid., 64–66. His plans included a major section on the head and some illustrations had already been drawn. For later insertions to the draft sent to Basle, see below, p. 101.

231. Vesalius, *Fabrica,* 538; *Fabrica* 1555, 662; *Fabrica* 1543, 510; 650, with O'Malley's discussion, *Andreas Vesalius*, 113, 436. His constant requests to the authorities for more human bodies to dissect were not always successful, Vesalius, *Epistola Chynae*, 194.
232. Vesalius, *Epistola Chynae*, 149.

CHAPTER 3

The 1543 *Fabrica*

Preparing the *Fabrica*: Artists, Printers and New Audiences

How much of a hurry is often underestimated.[1] The *Fabrica* is of a size and complexity greater than any previous book on the subject. A large folio, approximately 43 cm in height, and, excluding the index, over 660 pages in length, it contains an enormous number of woodcut illustrations. Each of the estimated 277 images had to be drawn and cut individually, even if some of the blocks themselves would later be set together on the same page to show the same structures from different angles or, as with the spinal vertebrae, at different levels. Nonetheless, by mid-August 1542, the wood blocks were ready for dispatch from Venice, where they were cut, and almost all the text would have been written by August 1 or soon after, the date given in the dedication.[2] The text itself was well over 300,000 words long, a remarkable achievement in little more than two and a half years of writing.[3] In addition, Vesalius had devised and organised a complicated system of references and fonts that had to be carefully explained to Oporinus when the blocks were about to be sent.[4] Vesalius' copy text that was sent to Basle was marked up with indications of where the illustrations should go and the type of font that should be used. A book of such size and complexity is extremely impressive even today and made possible, so Niccolò Massa later claimed, only by the

V. Nutton, *Andreas Vesalius and his Fabrica, 1537–1564*, Palgrave Studies in Medieval and Early Modern Medicine,
https://doi.org/10.1007/978-3-031-69565-0_3

relative absence of other duties and distractions such as those that filled much of his years in imperial service after 1543.[5] Decisions about what to include must have been made at an early stage, for creating images took time. All would have involved one or more preparatory drawings, and once made to Vesalius' satisfaction, even more time would have been required for the images to be transferred to the blocks and then carefully cut. The process would not have been speedy, even with the involvement of several artists and block-cutters.It is perhaps not surprising that they became bad tempered at being made to work so quickly and so long by such a demanding client.

Who these unnamed artists were has exercised Vesalian scholars for generations. The length of time required to produce the images, as well as the different styles of drawing that can be seen, sometimes, as in the musclemen figures, even within the image itself, excludes any notion of 'the' artist of the *Fabrica*. Rinaldo Canalis divided the plates into three groups: some schematic drawings, a large number of small and medium-sized figures of isolated or compound anatomical structures, and, thirdly, the famous 'musclemen' and the three skeletons.[6] The witty initial letters that open each chapter and depict putti and sometimes humans engaged in creepily playful activity as dissectors or surgeons may also have formed a distinct category of work. This classification is helpful, but only up to a point, for an artist could very well have worked on more than one type of image, and the identification of the block-cutters still remains mysterious.

Vesalius, who used schematic images in his lectures and in the *China Root Letter*, contributed to those in the first category himself as well as giving advice and criticism throughout.[7] Sometimes in describing the images he used the first person singular 'I', but elsewhere, when he employed the plural 'we', his personal involvement is less certain, although not excluded. In Book II, Chapter 26, he sketched a geometrical diagram of the outline of the muscle moving the scapula to clarify what had been depicted earlier, as well as, underneath, another diagram explaining what Galen had meant "obscurely" when he compared it to a trapezium.[8] He likewise included his own reconstruction of what the vena cava would look like if Galen had been correct in his oft-repeated statement that it split into two at the right side of the heart, inviting his readers to compare this with the true picture at Book VI, Fig. 5.[9] Later, in commenting on a small drawing of the seminal veins, he explains that "we decided to depict an alternative pattern" that is seen only rarely but which he then describes verbally.[10] A rough sketch of what the right and left cerebral ventricles look like from the side on is also likely to be his.[11] A further illustration in the notes to Book IV specifies how the schematic picture of the plexus of nerves above the first rib had come about: in his last dissection at Padua he had cut away all branches of the fifth, sixth, seventh, eighth and ninth pairs of nerves going to the arm of the corpse from which the illustration was made. He also warned that the plexus did not always display the same appearance, a useful way of deflecting criticism.[12] Other parts of the vena cava system, for example, the depiction of the fibres in the veins, he had earlier drawn for use in his teaching.[13] Those showing minor features of the vena cava system seem to have been prepared early and had their initial identifiers added later to coincide with those on the full length figure.[14] He may also have contributed to drawing some diagrams and small figures, such as the oblique line in the margin of p. 69 or the picture of two thoracic vertebrae from the body of a child that shows their ligaments, something that otherwise would have been difficult to reproduce without having attended the actual dissection.[15]

Vesalius in 1538 had been eager to secure the services of Jan van Calcar in any future anatomical project, and it is almost certain that Calcar contributed to the *Fabrica*, even if only briefly. Several scholars have pointed to representations elsewhere that suggest his participation in the *Fabrica*, even though they are unsigned and undated and of a lower quality than the final images in the book.[16] The strongest candidate is a drawing of the skull, femur and other bones, now in the Crocker Art Museum, Sacramento, that was drawn on paper widely used in Venice in the period 1542–1544. Interestingly, it shows studies of a jawbone and of the acetabulum of the femoral epiphysis that does not appear in the book, and at least one other representation contains superior details in its modelling, which together exclude that it was a later copy. Rather, it is, in all likelihood, a preliminary version from which Vesalius made his choice.[17] In 1970 Martin Kemp drew attention to a study of an écorché, then in a private collection in Munich, and suggested that it was a preparatory sketch for the image on p. 174.[18] A further image, showing the skeleton on p. 165, was identified in 1933 as an autograph by Calcar, but has since disappeared. All three show strong affinities with the images in the book, and, unlike many of those in category 1, are clearly the work of an accomplished artist. Since Vesalius began his book with what he knew best, bones and muscles, these images will have been among the first to be commissioned, a further argument for associating them with his friend Calcar, even if they all underwent considerable change before they appeared in print.

What has been claimed to be another preliminary drawing for the philosophical skeleton on p. 164 and attributed to Calcar was recently advertised for sale in Milan.[19] But there are strong grounds for believing that this is not by Calcar for, crucially, the identifying letters, which are claimed to be contemporary with the drawing and in the hand of Vesalius, are neither those of 1543 nor placed in exactly the same position.[20] They do, however, correspond exactly to the third skeleton in a well-known later book on anatomy for artists, François Tortebat's *Abregé d'anatomie, accommodé aux arts de peinture et de sculpture*, published in 1668, which has engravings after the Vesalian illustrations (Fig. 4.10).[21] The lettering in the drawing, particularly the 'R', resembles very closely that in Tortebat, and the sequence of initials is the same. Given these discrepancies, it is hard to escape the conclusion that this drawing and its annotations are not evidence for the early stages of the *Fabrica* but are connected, for whatever reason, with Tortebat.[22]

Another recent discovery appears, at first sight, to add further weight to the idea that the three skeletons, some of the images of bones and, perhaps, some or all of the musclemen, were at least in part the work of Calcar. In a copy of the 1543 *Fabrica*, once in the library of Elector Johann Friedrich of Saxony (d. 1547) and from 1549 until today in the Jena University Library, an unknown reader, possibly a doctor, commented on two of the illustrations.[23] Below the first skeleton man on p. 163 and again below the muscle man on p. 193 he added a brief note: *Joh. De Calchar Pictor Belgicus celeberrimus fecit Venetiis A. 1540* (Jan van Calcar, the very famous Belgian painter, produced this at Venice in 1540). This is an unusual comment, for not everyone would have considered Calcar a famous painter, even if they had read their Vasari, and the date of 1540 does not correspond to that of the *Tabulae*. It would, however, fit perfectly with what has already been suggested about his role in the *Fabrica*, as well as with an anonymous inscription on the reverse of a drawing of the title page, now in the Huntington Library in San Marino, CA, which also gives the same name, place and date.[24] But this is far from conclusive. The Jena annotator is likely to have lived a century or more later, and his addition of the place and date, like that on the Huntington drawing, could have been a shrewd guess based on what Vesalius himself said.[25] But even if Calcar drew the skeletons, the cooperative nature of artists' workshops makes it likely that another artist was engaged later to draw in the background landscape here and in the connected series of musclemen (Figs. 3.1, 4.10 and 4.19), where they form a continuous frieze representing the hills behind Padua.[26]

Calcar has often also been considered a strong candidate for the designer of the famous portrait of Vesalius that opens the *Fabrica*[27] (Fig. 3.4). Although there are problems of perspective in this image, including the over-large head, the fact that Vesalius himself commissioned the wood blocks and reused the same image for the Latin and German versions of the *Epitome*, the *China Root Letter* and the second edition of the *Fabrica* suggests that as a critical observer he saw nothing unusual about it. At least one other portrait has been attributed to Calcar from this period, and his friendship with Vesalius would have made him a good choice, but he is not the only artist who will have known Vesalius.

Whether, as the San Marino inscription suggests, above, p. 86, Calcar was also the designer of the title page is more difficult to determine (Fig. 3.3). Nothing in his work indicates that he was capable of such complex massing of figures within an architectural framework, some with an iconographical significance.[28] Such expertise is typical of an artist used to creating large religious paintings with a variety of individuals displaying different emotions or attributes, which is not true of Calcar. Besides, if the title page itself was among the latest to be designed, Calcar may already have left for Naples, although his presence there was not documented until early 1545. But even if the San Marino drawing may have been only a preliminary sketch rather than a later copy, the case against Calcar's authorship is a strong one, and it is not surprising that some have thought instead of Titian, with whom he had associated for a while.

This attribution was first made in 1668 by the French scholar Roger de Piles in Tortebat's anatomy book and was given wider publicity by the inclusion of the Vesalian images of the skeleton and the musclemen in the series of anatomical plates entitled *Notomie di Tiziano* by Domenico Bonaveri only a few years later.[29] Further evidence for Titian's involvement was put forward in 1943 by the Austro-American art historian Erika Tietze-Conrat, who claimed to have discovered evidence of Titian's involvement in a "notomia di Vercelli".[30] But one may doubt whether Titian at the height of his powers and busy with other compositions would have bothered to design a frontispiece for a book, however, importunate or wealthy its young author. Nor was he directly involved in dissection, for Patricia Simons and Monique Kornell have shown conclusively that the word 'notomia' does not here refer to any academic dissection by

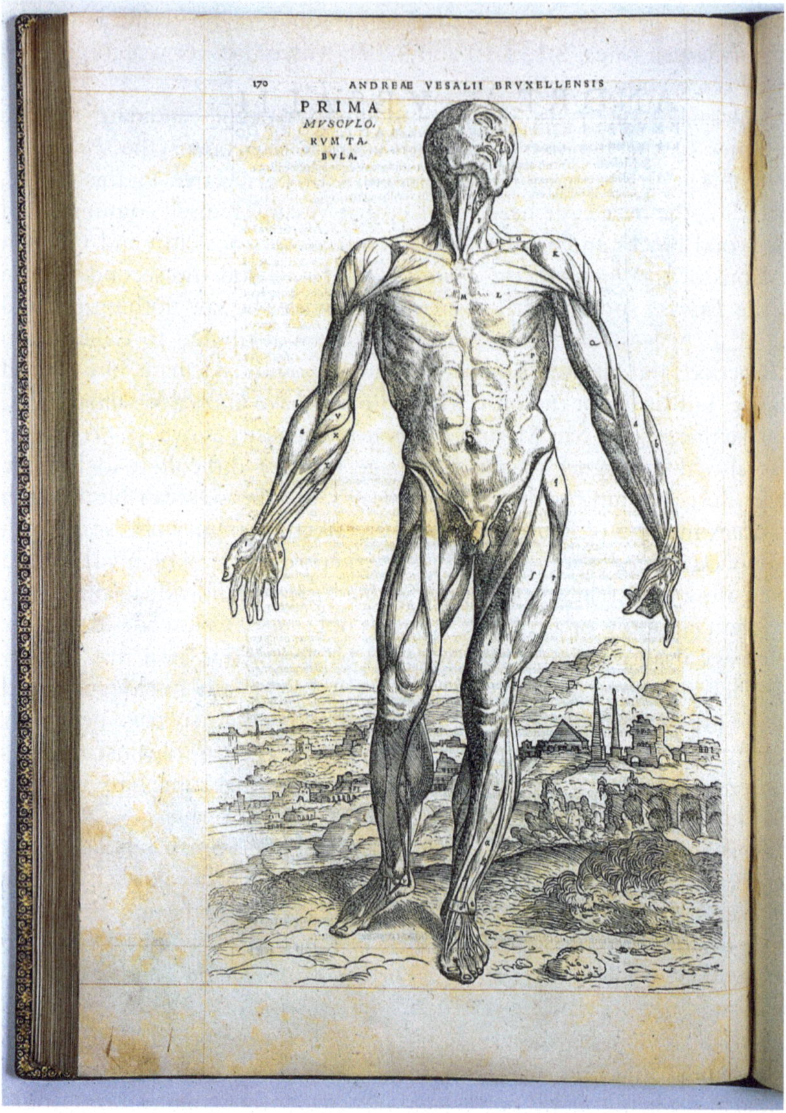

Fig. 3.1 Andreas Vesalius, *De humani corporis fabrica*, Basle: 1543, p, 180. The first muscle man. See also Figs. 4.10 and 4.19. (*Source* Wellcome Collection)

Vesalius but to the public drawing and quartering of a celebrated criminal, Battista da Vercelli, who was allegedly involved in a plot to kill Pope Leo X in 1517. Chronology also excludes any link between Titian and Vesalius at this point, for the reference comes from a speech given before 1539 and probably in 1536, long before Vesalius moved to Italy. Besides, given Vesalius' taste for self-publicity, he was hardly likely to have remained silent over any involvement and encouragement from the greatest artist of the day in Venice.[31]

Titian aside, a favourite candidate for the title page is Domenico Campagnola (1500–1564), another pupil of Titian, who painted several crowd scenes in the style of his master, although he is more famous today for his landscapes.[32] He was also a more than competent block-cutter, and Muraro and Rosand in their study of Venetian woodcut illustrations attributed the 'fanciful' initials to him for that reason.[33] He was also working in Padua and its environs at this time, so he might well have encountered Vesalius.

More recently, Martin Kemp in an unpublished lecture has revised his earlier support for Campagnola, and drawn attention to another young artist who is known to have been active in Padua in 1541–1542.[34] Giuseppe Porta, later known as Giuseppe Salviati (c. 1520–1573), became a pupil of the former Florentine painter Francesco Salviati in Rome in 1535, and accompanied him to Venice in 1539, even adopting his surname.[35] His crowd scenes are often complex, highly symbolic in the actions portrayed in them, and show individuals standing and seated on the ground. In this he is firmly in the Florentine and Roman style of Michelangelo and Titian, with whom he associated in Venice. When his master left in 1541, he remained in the Veneto to work, especially at first in the area around Padua. He was later responsible for many frescoes in palazzi in Venice as well as complex altarpieces, like the *Deposition from the Cross* in the church of San Pietro Martire in Murano.[36] He first made his name in 1540 with an elaborate frontispiece to a book on fortune-telling, *Le Sorti,* by Francesco Marcolini, a publisher in Venice, which depicts a group of men and women in a pastoral scene engaged in discussion together.[37] Although their numbers are fewer than on the *Fabrica* title page, the variety of the poses depicted is similar. Salviati also designed the portrait frontispiece and the small images that grace the fortune-telling cards shown in the book, which resemble, at least in size and in style, some of the witty initials that open chapters in the *Fabrica*. He continued to work for Marcolini for some time, including on the 1541 printing of Aretino's *La Vita di Catherina Vergine*.[38]

He was certainly in Padua in 1541, when plans for the *Fabrica* were already well advanced, for in the preface to a book of 1552, he records a meeting there with Daniele Barbaro "eleven years ago".[39] He may very well have met Vesalius then, for the anatomy scene on the title page of the *Fabrica* contains precise details that could only have been gleaned from attending a formal dissection, and it is tempting to suppose that Salviati was among the onlookers at one of Vesalius' anatomical displays either in 1541–1542 or earlier.[40] Alas, the attribution to him of a series of paintings from these years representing the life of St. John the Baptist in the then Palazzo Frigimelica, the family home of Vesalius' colleague in the medical faculty, Francesco Frigimelica, can no longer be maintained as they were originally painted in 1552 by Campagnola for the organ case in the church of San Giovanni da Verdara and were transferred to the palazzo only after some decades.[41]

In 1969, Francisco Guerra put forward another argument in favour of Salviati, suggesting that Marcolini was responsible for organising the images that Vesalius wanted for his book, and that the superintendent of his engravers, Johann Britt, received a large payment for his services sufficient for him to set up as a printer on his own.[42] Certainly, the gap in Marcolini's known publications under his own name coincides with the probable production schedule of the *Fabrica*, and Salviati's close association with Marcolini, if he was indeed responsible for designing the blocks, further strengthens his candidature. In short, like Calcar in 1538, Salviati was a young man of talent looking for a patron and an obvious catch for Vesalius.

But, whoever the artist of the title page was, it is clear that more than one hand was involved in producing the images in the *Fabrica*, and that Vesalius himself played an active role in choosing what to include and in supervising their preparation. He may not have been satisfied with the behaviour of those he employed but the result was vastly different from anything that had appeared before. Although, as Frank Hieronymus pointed out, there are precedents for all the individual developments introduced by Vesalius and Oporinus, it is their combination that sets the *Fabrica* apart.[43] In size, content, layout and complexity, to say nothing of the message it conveyed, it set new standards of accuracy and workmanship for books of all types, let alone those on medicine.

How unusual the *Fabrica* was can best be grasped by some statistics. At 43 by 29 cm, it is larger in its page dimensions than any other medical book, save the *Tabulae,* and, his own *Epitome* aside, it was not rivalled in size until the later editions of Mattioli's herbal in the late 1560s.[44] No other sixteenth-century book on anatomy comes close to it: Estienne's substantial treatises are some seven or eight cm. smaller in height. With some 663 pages of text, excluding the index, it is surpassed in length only by the 896 pages of Fuchs' *De historia stirpium*, which is printed in a slightly smaller folio format with the images of plants occupying roughly only half the book, and, later, by the 691 pages of Conrad Gessner's *Bibliotheca, Library,* of 1571. Each complete page contains on average 57 lines of type and roughly twelve words a line, written within three years at most, and requiring some six months or more to be set up on several presses by different groups of compositors.[45] Not only were there illustrations to be included, many times more than in any other book save that of Fuchs, some within the text, others occupying the full page, but there were also innumerable notes to be set in the margin as well as cross-references in both margin and text. Vesalius was well aware of the problems that this complex arrangement would cause, for he felt obliged to inform Oporinus in a letter accompanying the illustration blocks to Basle how he expected his marginalia to be printed. The outer margins were to be reserved for subheadings to the subject matter in the text, the inner for an elaborate system of references to the guide letters in the images, keyed in by superscript letters to the text. The text itself was not to be interrupted by the references themselves but was to be printed as a piece of continuous prose, even if it is too long and too dense for modern taste.[46]

Vesalius was also sensitive to the aesthetics of printing.[47] The wood blocks were cut on pear wood perhaps specially chosen to give the surface greater equality of texture and elasticity. Willy Wiegand, an expert printer and the publisher of the 1932 reprinting of the images from the original blocks, described how the cutter was able to achieve effects that enhance the beauty of the volume:

"The contour lines both swell and diminish. The parallel shading lines are remarkably long and model the shape of the figure by their more or less rounded effect. The deeper shadows are produced by crosshatching, which permits a strong contrast...The fine shading lines do not end abruptly but in a downward-cut curve on the face of the block, so that in the imprints these lines taper toward the end, giving a soft effect instead of an abrupt termination....Another peculiarity is that the parallel shading lines are not generally made by two cuts at right angles to the face of the block but with an oblique undercutting so that they look like a half-opened fan. Perhaps the hand of the woodcutter attained a gentler and more delicate movement in cutting in this way. The result is that, in printing, these fine shadow lines are somewhat elastic and give way slightly with the pressure".[48] The same techniques were used by other block-makers, not all of whom were as obviously skilled as Vesalius' workmen.

Vesalius was particularly concerned that the illustrations should be printed as clearly as possible, and that their identifying letters should not be overwhelmed by their surrounding context.[49] Similarly, he insisted on a variety of typefaces and sizes to distinguish between sections that referred solely to the images, printed in italics and keyed to the letters in the images, and those in the rest of the book. The sections in italics are printed in a slightly larger font than the roman and with a slightly wider spacing, which poses additional problems for the setter, but the result makes the italics as clear as the roman, and easier to read, while the slightly smaller number of words in the line and the increased length of the sections give them an equal emphasis.[50] The result, as Roger Gaskell explains, is two books in one; a continuous and unillustrated narrative text coupled with an anatomical atlas in which the text is a commentary on the illustrations. These latter sections function like Dürer's book on human proportions or Serlio's architectural treatises in which the text and images are interdependent.[51] Providing the cross-references, Vesalius explained to Oporinus, required just sheer hard work, not any specific skill, hardly consolation for a printer whose employees were charged with this complex setting.[52] It is perhaps not surprising that, as Dan Garrison points out, many of the marginal references in the final production are inaccurate because they were scribbled onto the proof sheets and were not entirely checked.[53]

Had the publication of Charles Estienne's treatise on anatomy not been delayed for years by a lawsuit, the extent to which he influenced Vesalius in integrating text and illustration, if at all, would have become clearer.[54] Estienne certainly attempted to use letters to refer his reader of the text to specific parts of the illustrations, particularly with the smaller ones in Book III, but often he contented himself with short descriptions placed in boxes within the woodcut images. But this was far inferior to what Vesalius was to achieve. Although many of Estienne's images of the body in his later books were drawn from actual dissections and were, for that reason, more accurate than earlier ones, they were cut into blocks that were then inserted into much larger ones, some of which had been modelled on a 1527 series of *Loves of the Gods* by Jacopo Caraglio that derived from pornographic poems by Aretino[55] (Fig. 3.2). The resulting detail is frequently too small to be portrayed accurately or to be discerned with the naked eye. Estienne's use of letters within the text of his exposition may have been the reason for Vesalius' complaint about this procedure, but others had long done this in non-medical treatises. But even if Estienne's plans had influenced Vesalius, the contrast between the two books shows the enormous gulf between them. Estienne accepted the constraints imposed on him by the printer, whereas Vesalius used all the resources of printing to provide an appropriate "pictorial quality" for his images. His understanding of the importance of shading, the use of different typefaces and the thickness of lines to give greater elegance shows his artistic sensibility as well as his familiarity with the most up to date developments in printing.[56] Even if he graciously yields to Oporinus' expertise and painstaking accuracy, his letter makes it clear how much he wished to remain in control.[57]

Vesalius had used a local printer for the *Tabulae Sex*, and it might have been expected that he would have used one of the more celebrated Venetian printers for his great work, given their expertise in all aspects of printing and the quality of their artists. Instead, he chose Oporinus and Basle, although he had the illustrations drawn and cut in Venice. He justified his decision in his preface. He appreciated the "divinely and wondrously talented" Italians, but complained about the weakness of the copyright provision that had allowed publishers from North of the Alps, especially in Germany, to plagiarise the *Tabulae*.[58] Basle, within the boundaries of the Holy Roman Empire, would not allow copyright privileges granted by the Emperor to be violated without penalty, and he expected his mother in Brussels to be able to send on an actual privilege document that he had been given some years before that would cover all his publications.[59] But this cannot be the sole explanation, nor can the good relations he had established with Robert Winter with the publication of the revised *Paraphrasis* and the 1539 letter on bloodletting. Given the size and complexity of the production of the book, a search for a potential publisher must have begun by then at the latest, for Vesalius required some assurance that his plans were capable of being fulfilled. There must have been several letters exchanged between Winter,

Fig. 3.2 Charles Estienne, *De dissectione partium corporis human*, Paris: S. de Colines: 1545. p. 276. The child in the womb. The outline of the inserted block can be seen clearly. (*Source* Wellcome Collection)

Fig. 3.3 Andreas Vesalius, *De humani corporis fabrica*, Basle: 1543, title page. (*Source* Wellcome Collection)

Fig. 3.4 Andreas Vesalius, *De humani corporis fabrica*, Basle: 1543, frontispiece. (*Source* Wellcome Collection)

Oporinus and Vesalius, and perhaps others, before the agreement was reached.

But in 1541, Basle was still not an obvious choice for a work on this scale.[60] True, publishers like Froben had over the previous two decades established a reputation for their scholarly publications, but it was only very recently that they had become a major rival of the Italians in every respect. The edition of the collected works of Hippocrates in Greek put out by Froben and Episcopius in 1538 evidenced the quality of their printing and the scholarship behind it.[61] The same year saw the huge Basle edition of the complete works of Galen in Greek that involved substantial co-operation within and without Basle. It was edited by Joachim Camerarius (1500–1574), professor of Greek at the University of Tübingen, assisted by Leonhard Fuchs, his colleague there, and by Hieronymus Gemusaeus (1505–1544), professor of Physic at Basle and a distinguished Hellenist. Important suggestions for emendations were sent to Basle by the English Galenist, John Clement (1500–1572), who had been one of the editors of the 1525 Aldine edition of Galen. Its production was made possible by a contract between five of the leading Basle printers, Andreas Cratander, Johann Bebel, Michael Isingrin, Johann Herwagen and Johann Erasmus Froben. This edition was not only superior to the Aldine in the quality of the Greek text it contained, but it also showed that Basle could now publish large and expensive volumes to challenge the Venetians.[62]

It may have been the success of this collaborative project that encouraged Isingrin and Fuchs to embark upon the most lavishly illustrated medical book so far envisaged, Fuchs' *De historia stirpium*, *Investigation into Plants*, which made Isingrin's name as a printer. Although it did not appear until 1542, the plans of Fuchs and Isingrin must have been known in the small world of Basle well before then.[63] It is likely, although far from provable, that their example stimulated Oporinus, a similarly young man of ambition, to leave his job as a teacher of Greek at the university to take over the business of his brother-in-law and commit himself full time to printing and publishing.[64] What Isingrin planned to do for medical botany, Oporinus could do for anatomy.

Although he had been part of the consortium with Winter, Platter and Lasius that had bought up the press of Cratander, he seems at first not to have been involved in its day-to-day workings. But he had grown worried about Winter's financial abilities and was planning at least as early as July 1541 to branch out on his own.[65] The first book from his own press appeared only in November 1542, but it is clear that right from the start he intended to make his name with some spectacular and controversial publications. His 1542 plan to publish an edition of the *Koran* in Latin aroused the wrath of the city council, and it was only through the intervention of the Swiss scholar Joachim Vadianus and the theologian

Martin Luther that it allowed this *succès de scandale* to be issued in an unsigned edition with a preface refuting it by Melanchthon and with its sale banned within Basle itself.[66] The example of Isingrin not only reassured Oporinus that in Basle enough funds could be found to subsidise the production process, but also provided him with a model of typographical excellence that could be imitated, if not surpassed. But it was still a risk, for Oporinus had never been involved in publishing a volume that contained illustrations or one of such size. Indeed, as Martin Steinmann points out, the *Fabrica* aside, his productions were mainly of plain texts in octavo format.[67] Vesalius might praise Oporinus as a very careful printer, far superior to those who had plagiarised the *Tabulae Sex*, but he was still concerned that the wooden blocks might come to some harm during the production process, and thus be incapable of being reused, should Vesalius require them.[68]

Printing in the Renaissance was always financially precarious, and the *Fabrica* was an extremely expensive book to produce, even if Vesalius himself paid for the actual wood blocks.[69] Sachiko Kusukawa, basing herself on the later figures from the Plantin press, suggests that an illustrated book might cost four times as much as one without illustrations. Paper had to be of the best quality, although it is not easy to determine how good that used for the 1543 edition was. Some of the pages I have handled are relatively thin, particularly when they are in a poor state of conservation, but others are beautifully made. But undoubtedly Vesalius expected Oporinus to provide paper whose denseness and smoothness would match those he knew in Venice and which he had used for the proof of the illustrations he sent to Basle.[70] The sheer size of the page was also pushing the boundaries of paper technology and would also have involved both the paper-maker and the publisher in greater costs than a normal folio (perhaps two to three times those of paper in octavo size). Three years later, Vesalius acknowledged this by offering to pay Oporinus for the extra expense involved in printing his *China Root Letter* on the best quality paper with wide margins to make the setting easier and give greater legibility.[71] The large folio *Fabrica* would have required an even greater outlay on paper, even if transport costs could be reduced by having it, as seems likely, produced in the paper mills of Basle.[72] Producing a book like the *Fabrica* was a very expensive risk.[73]

How expensive is impossible to quantify with any degree of precision. Nonetheless, there are certain indications that allow some judgement to be made. Within a few days of publication, a friend of Vesalius, Johannes

Gast, reported that in Basle the cost of the *Fabrica* alone was four florins and four and a half *batzen*, and of the *Epitome* ten *batzen*.[74] This is roughly in line with what Oporinus was charging in 1547, five florins and three *batzen*. Sachiko Kusukawa equates the cost of the *Fabrica* at thirty-nine meat meals or thirty-four fish meals at a Basle inn, but a better comparison would be with university salaries, for academics would be among the hoped-for purchasers. A Basle florin was the equivalent of two Basle *pfund*, so the *Fabrica* by itself would cost nine *pfund*. One of the two Basle medical professors at the time was paid 75 pf. a year, the other and more senior 100 pf. Alban Thorer, who lectured in the medical faculty and later became the translator of the *Epitome*, was paid a mere 40.[75] Elsewhere a copy in Antwerp cost 4 gold guilders (plus one for binding), and the copy that was said once to have been owned by Cardano cost six florins. The 1555 edition, being larger, may have cost a little more: a copy owned by a Cracow humanist and poet cost 8 florins (and 32 groschen for binding).[76]

It is not easy to determine for certain how many copies were printed and estimates range widely. Margócsy and his colleagues, following the model of Owen Gingerich's studies of surviving copies of Copernicus, estimate that at least 600 copies were printed of the first edition.[77] If all of them were sold, save for those Vesalius gave to friends, which seems likely, and assuming that Oporinus made a profit of around thirty per cent, his production costs would have amounted to some 3600 pf., giving him a tidy profit of 1800 pf. But these are mere estimates: more significant is the evidence that this was an expensive book, but one not beyond the pocket of elite doctors or those practising in larger towns.[78]

On August 24, 1542, Vesalius wrote a detailed letter to Oporinus that contained information about the progress of the work. The blocks had been finished, packed carefully with the help of the block-cutter and a young friend, Nicholas Stopius, to avoid damage, and handed over to a firm of Milanese merchants, the Danoni, for transport across the Alps. He announced that when he finally came to Basle, he would bring with him the decree of the Venetian Senate forbidding anyone to print the book there without its authorisation, and his mother in Brussels would forward a similar privilege from the Emperor. Oporinus had already obtained the printing privilege from the city of Basle, and Vesalius hoped that he would be able to get a similar one from Francis I of France through the intervention of the Bishop of Montpellier, the French ambassador to Venice. But Vesalius was not too worried if that did not arrive, for he did not wish to clutter up a page with official documents that unscrupulous booksellers around Europe happily disregarded.[79] He had decided to dedicate this volume to the Holy Roman Emperor, Charles V, of whose interest in matters anatomical he had heard from his father, the imperial apothecary, who had earlier shown him a copy of the *Tabulae Sex*. It was, as Sachiko Kusukawa says, an elaborate job application as well as a medical treatise.[80]

The letter of dedication combines appropriate flattery and a description of the book's contents with a tendentious account of the history of anatomy. In it Vesalius paints himself as the heir to a classical Alexandrian tradition of human anatomy established by Herophilus and Erasistratus in the early third century BCE and continued down to Marinus and Galen's contemporary and opponent Lycus of Macedon in the second century CE.[81] This, he claims, had long been forgotten except for a few fragments as a result of the laziness of the medical fraternity in following Galen and in cobbling together incompetent summaries of his writings.[82] By following Vesalius, students who dissect properly will thereby be returning to the truths of 'the ancients'.[83] Galen is praised for his work in therapeutics and surgery, but although he is deservedly called the first among the professors of dissection, he had never dissected a human body, unlike his predecessors, and it is not surprising that his work contained many mistakes. This was a claim that Vesalius knew would arouse accusations of impiety among partisans of Galen, and which he sought to rebut in two ways.[84] First, he praised Galen as an experimentalist willing to challenge authority where he thought fit, and, secondly, he pointed out that even his strongest supporters had noticed discrepancies in his writings and had encouraged others to investigate them. This, he averred, was the way forward: greater acquaintance with the dissected human body would lead to the renewal of true anatomy. The medical Guild would thus both be reformed and, at the same time, retain strong links with its distinguished past.[85] It is a masterpiece of rhetoric and elegant Latin.

Having obtained the university's permission to go on leave, Vesalius departed for Basle in 1542 to supervise the printing of the great book. When he reached there is unknown. There are indications that he spent some time with the anatomist Canani in Ferrara, where his brother, Franciscus, was also studying, for he saw at least some of the pictures of the muscles of the hand that Canani was planning to include in his forthcoming book.[86] His own letter to Oporinus at the end of August does not suggest that his departure was imminent.[87] He had certainly reached Basle by January 1543, for he appears in a matriculation list from that month and Padua appointed Realdo Colombo on January 19, 1543, to be his substitute.[88]

The next few months were involved in seeing the book through the press. Vesalius was involved in proofreading and in making additions and corrections. Some introduced new information. The young lawyer Christopher Pfluegel of Salzburg, who had earlier told him about the strangely shaped skulls of the inhabitants of Styria, is now additionally praised for his role in ending the second siege of Louvain in July 1543.[89] Even before his arrival in Basle, Vesalius had already been forced to write to Oporinus to apologise for an error in the first plate of the nerves on p. 332/432.[90] The letters l and m had been originally left out, by inadvertence either on his own part or on that of the block-cutter who had failed to notice them. He realised that introducing these letters would possibly cause more damage than was warranted, and instead sent a small drawing of the relevant pair of nerves that could easily be printed separately. Oporinus, or his setter, followed his advice only too well, printing in the notes to the image both Vesalius' explanation of the image and his instructions to Oporinus.[91]

A few pages earlier, at p. 329/429 (Fig. 3.13), there is in the margin what appears to be a curious afterthought. At a late stage in the production, to help the reader understand what Galen in his discussions of the recurrent nerves and elsewhere had meant by the word '*glossocomion*', 'box-splint', Vesalius asked Oporinus to cut two replacement historiated initials (for F and E) in order to show the apparatus being used in reducing a dislocation of the femur.[92] The word itself is not to be found on this or the subsequent page, but only in Galen, for the relevant passages are not cited or given a precise reference in the *Fabrica*. But Vesalius had found the instrument useful in dealing with dislocations of the arm as well as of the femur, referring his readers for a further discussion to a future book and another context (perhaps dealing with surgery). Since this was a project that he soon abandoned, he removed these comments and the illustrations from the second edition. Not surprisingly in such a complex piece of setting, some errors still remained: two small figures were printed upside down, the page numbers in the middle, from pp. 213 to 391, are out by a hundred, and, most obvious of all, the fourth figure of the 'muscle man' is presented as the first. But Vesalius was not entirely occupied at the press. In May 1543, a notorious criminal, Jacob Karrer, was beheaded and handed over to Vesalius to be dissected, and his skeleton, once mounted, was to be passed to the university to be kept there, a friend reported, as a "constant memorial" of his visit. It can still be seen today in the University Anatomical Museum along with another skeleton prepared in 1573 by another famous Basle anatomist, the slightly later Felix Platter (1536–1614).[93]

Vesalius was also involved in two other directly related projects. The first, his *Epitome* of the *Fabrica*, can be considered a (far superior) revision of the *Tabulae sex* aimed at a student audience, as well as being an excellent digest of the main outlines of the great book.[94] It was printed on separate sheets and considerably larger in format than the *Fabrica*, some 57 × 29 cm, from which it borrowed the blocks for the title page, the portrait, the 'philosopher' skeleton, the two broadsheet inserts and several of the initials. Its organisation was very different from that of 1538, and Vesalius in his preface felt it necessary to explain how it should be used. The first six chapters, he wrote, contain a description of the body suited to beginners, but with its parts explained in a new order, just as he had done in his lectures in Padua and Bologna.[95] The first pair, on bones and muscles, reproduces that of the *Fabrica*, but they are then followed by chapters on the digestive system, the cardiovascular system, the nervous system and the brain, and finally, the reproductive system. The reader is then to turn to the illustrations, although others, he acknowledged, would begin with them. The male and female figures in the middle should be viewed first, for they show the surface of the body, and are accompanied by an index of the Greek terms and the new humanistic Latin ones. The student is then to move backwards, beginning with the skeleton of the 'philosopher', and then proceed in reverse through the five plates of the muscles. Space is saved by having the skeletons display various aspects of the dissected skull and brain and by placing smaller structures at the base of the figures. All the illustrations are keyed into a brief explanatory text. They are also designed so that some can be cut out to form flap anatomies, one of the male, the other of the female, by attaching the individual images from the last three figures in the book, which show the blood vessels, the nerves and the viscera on to a base skeleton. (The first two were also used in the separate broadsheet inserts in the *Fabrica*.) Vesalius then explained in detail how these figures should be cut out, mounted and then carefully glued in place to make a flap anatomy. This was a relatively new innovation, used by several of the precursors and imitators of his *Tabulae sex* to create their own simple fugitive sheets. If done correctly—and that was not always the case—the student could visualise the body in three dimensions.[96] As an illustrated student guide, this was a marked improvement over anything that had gone before, and, as Vesalius explained in the *Fabrica*, the fact that its production began a little later than that of the big book meant that, in depicting the nervous system at least, it was possible to be even more accurate. It succeeded in showing some of his own discoveries in a cheap and accessible form, and, propped up on a chair, it could be used as a guide during a dissection lesson. That only a small number of copies have survived today suggests that it was often heavily used and simply fell apart.

The text and the anatomical woodcuts of the *Epitome* were prepared in Italy at the same time as the *Fabrica* itself. Although its dedication bears a date twelve days after that of the *Fabrica,* it was printed in Basle in June 1543, a few days before the larger book.[97] Its dedicatee was the later Philip II, the son of the Emperor, who, like his father, was informed by Vesalius of its author's role in the restoration of anatomy after centuries of neglect.[98] Both presentation copies were lavishly printed on vellum and further decorated in glorious colours, in which the imperial purple is prominent. Both attest not only the regal magnificence of the dedicatees but also the wealth and status of the author who presented it.

If the *Epitome* can be seen as a continuation of the *Tabulae sex*, his other project broke new ground, a translation of it into a vernacular language, German. Several anatomical fugitive sheets had already appeared in German, Flemish, French and English, but the *Epitome* was the first such academic anatomical treatise and on a larger scale than anything previously printed.[99] The translator was Alban Thorer, (1489–1549/50), who had first taught Latin at the university before turning to medicine around 1528. He practised in Basle as well as teaching within the medical faculty from at least 1536 although he never held one of the two chairs in medicine. From 1535 he was also a doctor to the Margrave of Baden and became rector of the university for a year in 1542. Although he had published a treatise on the plague in 1539, his major reputation was as a translator of the medical classics into Latin, including Paul of Aegina, Alexander of Tralles and Soranus of Ephesus.[100] In 1544, he published a paraphrase of Book X of the *Ad Almansorem*, following the same principles as Vesalius had done for Book IX.[101] Given the close relationship of both Vesalius and Oporinus to the medical faculty, it is highly likely that Vesalius approved of Thorer's work, even if the decision to arrange for a German version was not his own. He certainly stayed in Basle during the process of translation and printing, for he stood as godparent to the son of Robert Winter on August 3, less than a week before Thorer dedicated his preface to the Duke of Würtemberg.[102] Thorer's sentiments echo those of Vesalius: this is an "extremely useful" contribution to improving medical education for everyone in the "German nation", lay as well as professional, so that all could shun the empirics and quack doctors who were to be found everywhere. He appealed to the Duke to legislate against these charlatans, reminding him of classical precedents for a royal interest in anatomy.[103]

In translating Vesalius' Latin into German, Thorer was appealing to a new and different audience. Relatively few, even among educated males, had enough Latin, or the stamina, to cope with Vesalius' ornate Latin, and there was a much wider public among German speakers who constituted the urban middle class in towns from the Rhine across Europe into distant Transylvania. The *Epitome*, a compromise between the long and detailed exposition of the great book and the skimpy accounts in small medical handbooks and fugitive sheets, was ideal for retailing the latest information on the make-up of the body to an educated readership. But it was not without its problems for the translator. The new humanist anatomy was only just expanding beyond the medical school, so that Thorer had to create almost a new vocabulary in German in order to produce a version that was succinct, understandable and true to the original as well as elegant.[104] This was difficult, for Latin is a far more concise language than German, and, much as Thorer tried, his version took up many more pages than the original. But this gave Oporinus the chance to include yet more of Vesalius' images, one of which had to be recut in order to be captioned in German.[105]

The Message of the *Fabrica*: The Images

Vesalius expressed the message he wished to deliver to his reader both visually and verbally in a way that emphasised his own singularity and the significance of his project. He intended to make a break with the past as it was interpreted by contemporary anatomists while at the same time retaining the links with the classical heritage that was the sign of a man of culture and learning. This set him apart from others, mere 'mechanicals', such as surgeons, who involved themselves with blood and gore, while indicating his own intellectual and social standing. An author who could refer in passing to his family's close acquaintance with the greatest ruler of the day was no ordinary practitioner. His decision to address his publisher by an academic title that he no longer held further demonstrated that this was a work that would meet the highest standards of accurate scholarship, while the quality of his images, visible from the title page, emphasised his artistic connoisseurship as well as his and his publisher's ambition to produce something that was vastly different in scope and in potential readership from anything that had gone before.

The title page of the *Fabrica* is of a size and complexity rare even for a non-medical book (Fig. 3.3). Its elaborate combination of details that convey multiple meanings is a far cry from the simpler and more sober title pages of most books that contented themselves with a title, the name of the author and, increasingly, that of the printer/publisher as well. It portrays a single scene, of Vesalius demonstrating a dissected corpse, in a far more elaborate and artistically superior way than any earlier title page of a medical or scientific work.[106] It also differs from those in several editions of Galen published by Simon De Colines or in the Juntine editions of Galen and Hippocrates, which assemble smaller images together around the title information, some depicting specific incidents, others great figures in medicine from the past.[107] A few earlier frontispieces and illustrations of anatomy scenes include spectators in attendance. Six appear in the series of printings of the *Fasciculus Medicinae* showing a dissection scene, but not everyone appears to be interested in what is taking place.[108] There are more than a dozen in the frontispiece to De Colines' 1531 printing of Guinter's translation of Galen's *Anatomical Procedures.* Many of them wear academic dress as they attend to the words of a physician explaining what is being revealed by his dissector of the corpse before them.[109] In the *Fabrica*, by contrast, Vesalius' dissection takes place in front of an enormous number of onlookers, a clear indication of his reputation as a man who, on one occasion, attracted two hundred spectators or more to see him perform a dissection.[110] This crowded scene was intended from the beginning, as can be seen from three drawings preserved in Stockholm, Glasgow and San Marino that are generally taken to represent different stages in the evolution of the title page.[111] This is no static image, but an animated and carefully structured view of a varied and interested audience. Figures stand on the bases of columns or lean from a balcony to get a better view; they comment on what is before them, pointing and talking to each other. Even the figure at the foot of the illustration in classical dress turning away in disgust is responding to what is being said by pointing to the dissected body. Only a very few spectators appear to be distracted, either by animals or by the behaviour of those nearby. One student has brought along a large book, perhaps the *Tabulae sex* or, more likely from its shape, a book of Galen, which he is reading to check what he is hearing. The varied dress and different ages of the spectators, mostly the well-to-do or academics but also including a friar, two women (if only peering through the columns at the rear), and, if David Soulier is right, Marcantonio Contarini, the Venetian *podestà* of Padua, emphasise the ambitions of the author.[112] The size and behaviour of the audience convey an important message: this display is not just for students or medical men but has a wider relevance to society as a whole. Everyone, both old and young, man or woman, should be interested in the anatomy of the human body.[113]

It is a picture of action, from the gestures of the anatomist to the behaviour of the animals and their guardians. By contrast, the frontispiece of the *Fasciculus*, which Vesalius almost certainly knew, is dominated by books stacked in shelves or in cupboards, while in that to the Leipzig edition of Mondino the teacher sits with Mondino's book on his lap, just as in some versions of the anatomy scene in the *Fasciculus*. Here the three books, including that held by the student, are unobtrusive.

The setting is ambiguous, for it combines an actual event, generic characters and, in some interpretations, portrayals of known individuals in a situation that brings together disparate elements.[114] Formal anatomies at Padua or Bologna, for instance, took place in a temporary wooden structure, and the presence of foliage on the architrave of a column on the left might indicate a setting at least partly in the open.[115] Otherwise, the dissection is taking place in a temporary structure within a large, airy, building, with splendid Corinthian columns supporting an upper gallery resembling that in the semi-circular apse of a church. In a marginal note at the end of the book, in describing how best to conduct an anatomy, Vesalius reveals that the illustration represents half of the space with which he had been familiar in Bologna and Padua. Holding dissections in a church, often the largest available space in town, would have had the further advantage that the winter chill inside would have certainly delayed putrefaction at the expense of the audience's comfort.[116] This is also a place of spectacle, a true *theatron*, as indicated by the large theatrical mask beneath the cartouche containing the book's title.[117]

The reader's eye is drawn immediately to three larger figures dressed in ancient dress. One, deliberately turning away and looking at a dog, is likely to represent Galen, unable to face Vesalius' overthrow of his largely animal-based anatomy, something also indicated by the presence of a monkey or barbary ape, one of Galen's favourite subjects, at the opposite corner. The identification of the other two figures who stand at Vesalius' side is less certain: Andrew Cunningham suggests that they may be identified with either Aristotle, Herophilus or Erasistratus, and possibly the latter pair, who are singled out for praise in the preface to the *Fabrica*.[118] Like most of the audience, they are taking a great interest in the dissection, unlike the pair of figures quarrelling at the foot of the table. They are possibly surgeons or assistants, who would have performed the actual cutting before Vesalius' innovation and whose relationship to the now superseded Galen is indicated by the man in the hat, who is pointing to Galen with his finger. Other contemporaries have been identified with less plausibility on the basis of a likeness to their portraits or other descriptions. The bearded figure in the centre in Jewish dress may be Lazarus de Frigeis, a Jewish friend who helped Vesalius with his reading of Avicenna. Realdo Colombo, his successor at Padua, may be the man leaning over to touch the shoulder of the man trying to restrain the dog, while Oporinus may be the bearded figure looking down from the upper balcony on the right. There is no overt mention of him in the title page, unlike in the second edition, although his may be the initials I and O, that are fused together at the top of the illustration to form the Greek letter Φ, which can also signify 'philosophy'.[119]

The central space in this crowded room is taken up by the most unusual of all anatomical events, the dissection of a woman.[120] It is an image at one and the same time historical and symbolic. It is first and foremost a human corpse: the monkey, a favourite of Galen and the dog, regularly dissected in university anatomies even by Vesalius, are pushed to the margins of the picture as if no longer required. It is also the body of a woman. Public dissections of women were rare, partly because they were less involved in violent crime, partly because they were treated more leniently than male malefactors, and partly because most of the women who were convicted were local, and the authorities preferred to look elsewhere for corpses for fear of antagonising their families.[121] University statutes might specify an anatomy of a woman at least once every three years, and annually at Padua, but as a letter of the University of Pavia in 1464–1465 to the Duke of Milan made clear, circumstances might prevent them being performed for some considerable time.[122] This deficiency is also clear in artistic representations of the female organs. Leonardo da Vinci's drawings of the female anatomy and of the child in the womb are often fanciful and less accurate than those of the male, and although those of Vesalius are more accurate, they are among the weakest of his images.[123] Later in the *Fabrica*, he twice drew attention to a recent dissection of a woman who had failed in her attempt to avoid execution by pleading pregnancy, a plea that was rejected after she had been examined by midwives and interviewed by Vesalius.[124] Vesalius had her anatomy specially drawn to show the position of the bladder and uterus as well as the uterus in detail, and some of his readers might well have remembered the occasion.

By choosing a female body as the central object of his title page, Vesalius was not only announcing his own expertise in a way that visually tied in with his later account of the number of women he had dissected. He was also making a claim for anatomy to be able to reveal the most intimate secrets of the human body in a way that nothing else could. He is demonstrating an open womb, now deliberately made visible to most of his audience for possibly the first and even only time, and the site of the miracle of human creation. The first finger of his left hand pointing upwards to the heavens reminds his audience that the human body is the greatest work of the divine Craftsman, a trope that went back at least to Galen and was adopted by both Evangelicals and Catholics.[125] By also including a representation of an actively climbing naked male on the left, and, centrally, a large posed skeleton, articulated as if for a demonstration, Vesalius is thus interpreting a central feature of Christian theology. Whether in life or in death, unclothed, dissected or a mere assemblage of bones, the human body reveals the wisdom and forethought of its divine Craftsman.[126]

At the centre, gazing at the reader, stands Vesalius, elegantly and expensively dressed in a damask doublet, despite the bloody nature of the occasion. Unlike similar participants in earlier anatomy scenes, he stands alone explaining what can be seen and, by implication, what he has already done, for there is no visible role for either *lector* or *sector*, the two other main participants in the formal university anatomy.[127] The solitary reader of a book follows it for himself, head down, while the surgeons or assistants are squeezed out of sight of the audience beneath the dissection table. Two other figures to the right of the skeleton also carry a book, one open but scarcely glanced at, the other closed as its owner points to the corpse. The rhetoric of the image thus emphasises visually the main message of the *Fabrica*: a true understanding of the workings of the human body can be demonstrated only by an expert, learned, experienced and practised in dissecting by and for himself. The anatomical book is displaced as the prime source of knowledge by the evidence of the corpse itself.

That this title page shows truths about Vesalius as well as about the body is further indicated by the inclusion of Vesalius' family coat of arms, three weasels courant on a sable field, a pun on its origin at Wesel, whose civic coat of arms also features three weasels courant. The shield which dominates the top section is held up by naked putti, whose activities also feature throughout in the initial letters of many chapters.[128] The entablature above the columns also displays two other heraldic symbols associated with Vesalius, the Lion of Venice and the Ox, the 'Bo', of the University of Padua.[129] Mortices were cut in the two cartouches by the Venetian engraver, and the words were added in Basle by carving them on a separate piece of wood which was then glued into place.[130] The lower one records the copyright privileges from Venice and the King of France (the result of the Bishop of Montpellier's intercession), the upper, and much more prominent, the book's title.[131] Artistically, the latter plays a crucial role in the whole design, linking the two main levels of the image together and making the title itself clearer by isolating it from the activity going on above and below it and setting the words within a large blank space. As in the second edition of the *Institutiones*, Vesalius' name is printed in very large capitals, followed on the line below in smaller type by his place of origin (Brussels) and the first word, *SCHOLAE,* of his academic position, "Professor at the Padua medical school", before concluding in yet smaller letters with the rest of the description, followed by the title of the whole work, the *Fabrica* and its seven books.

The mention of his academic position at one of the most important European medical schools affirms both his own standing and the truth of what he is going to say about the *fabrica* of the body. What he intended by his choice of the word '*fabrica*' has been much discussed. The Latin word can mean 'fabric' in the modern English sense, but, although regularly so used by translators, it does not entirely fit a description of bones, arteries and organs.[132] Others have suggested that Vesalius was using the word as a Ciceronian allusion to mean the 'workshop' of the body, a living and working organism, a very common usage in classical Latin.[133] But how the body works and what it produces are secondary considerations in this book. It is more likely that, at least for the title. Vesalius was following the example of his colleague and one of the sponsors for his degree, Junius Paulus Crassus (c. 1500–1575), who had published in 1536 a Latin translation of an early Byzantine Greek anatomical treatise by Theophilus Protospatharius.[134] Written at some point between 700 and 900 CE and heavily based on Galen's *On the Usefulness of Parts,* it bore a title that can be translated into English as '*The Constitution*' or, better, '*The Construction*' or '*The Make-up of the Body*', and which Crassus turned into Latin as '*Fabrica*'. Given that the subject of Vesalius' book corresponds precisely to that of Theophilus and that he and Crassus were colleagues, this may be how he intended his readers to understand his title.[135] Recent English translators, however, such as Richardson and Garrison have kept 'Fabric' in the title even at the expense of possible misunderstanding, although they have preferred to retain the Latin word in their introductions and notes to avoid further ambiguity.

Unusually, the *Fabrica* begins with a second image of the author, this time drawn in close-up dissecting the muscles of the arm and hand[136] (Fig. 3.4). Although Vesalius was not the first anatomist to be portrayed in this way, for a similar picture of Achillini graces the title page of his anatomical book, the superior quality of this image reminds one immediately of the portrait of Fuchs that opens his *De historia stirpium*.[137] Fuchs, wearing an elegant coat, cape and cap, holds a sprig of germander speedwell, demonstrating in a single image his social, academic and intellectual qualifications for his theme. Vesalius is here similarly well dressed in a richly brocaded over-shirt patterned with leaves, hardly suitable, one might think, for the messy business of dissection. However, it would have been appropriate dress for a public disputation, as might be gathered from an incident in Florence only a few years later. In 1546, Jacopo Coppa, an irregular practitioner of some ability, faced with complaints from the surgeons of Florence, was given permission to make a public trial of his drugs in the main square. He appeared from his tent "in sublime majesty", wearing a black velvet tunic and skirt, a black cap and a gold chain, an indication that he was the social and intellectual equal of any physician or surgeon.[138] This portrait of Vesalius can also be compared with that, attributed to Francesco Salviati (1510–1563), of his pupil and rival, Realdo Colombo, recently identified in the collection at Tatton Park, Cheshire.[139] Colombo too wears a black doublet, although one much simpler in design, and he is seated at a table on which rests a large illustrated anatomy manuscript, probably his *De re anatomica*.[140] It is a conventional picture of an academic in his study and lacks the vigour of that of Vesalius, especially in its coloured copies, where Vesalius is represented as a man of action at the height of his powers.[141] He gazes firmly at the observer, while holding on to the arm of a somewhat exaggerated female figure that has been stripped of the skin from above the elbow to reveal the muscles of the lower arm and hand.[142]

The choice of this part of the body is a deliberate counterpoint to the opening of the womb on the title page, for both reveal the secrets of creation, albeit in different ways.[143] The anatomy of the hand is a different and more complex world than that of the womb, or, for that matter, the bowel or the vein. If the womb is a static receptacle, the hand is dedicated to movement, as tendons, nerves and blood vessels accommodate themselves to each other as the arm, hand and fingers perform their functions. It is at one and the same time powerful and vulnerable, simple and extraordinarily complex.[144] It is no wonder, then, that in a long tradition going back to Anaxagoras, a Greek thinker in the fifth century BCE, the human hand was considered one of the greatest wonders of divine creation. Anaxagoras had declared that it was through having hands that man was the most intelligent of beings, an observation Aristotle turned on its head to argue that it was precisely because he was the most intelligent of beings that Nature had provided man with a hand, the most versatile of all instruments.[145] A follower of Aristotle put it more succinctly. "God has provided us with two inner instruments with which we use outside instruments, the hand for the body and the intellect for the soul".[146] This theme was taken up at even greater length by Galen in his *On the Usefulness of Parts,* whose first book is largely dedicated to the wonderful work of the Creator or Nature in designing the human hand and arm. This message is repeated at length in Book II and forms part of the 'epode', the prose hymn to the Creator that concludes the treatise.[147] In one of the most familiar of his tracts, *On Mixtures*, one of the basic texts to be studied in Western universities, Galen praised the skin of the hand as the prime yardstick of the organ of touch and most suited to the most intelligent of all animals.[148] The Aldine edition had revealed other repetitions of this idea, not least in the tract *On the Opinions of Hippocrates and Plato*, where Galen had built on Plato's teleology as found particularly in the *Timaeus* to stress the purposefulness of the Creator. His praise of this "organ of organs" was repeated by other Renaissance anatomists. Vesalius in 1543 was not alone in describing the muscles of the hand and arm as the product of "the wondrous workmanship of the Creator". Niccolò Massa, in his 1536 tract on anatomy, had similarly called them "most beautiful" and "the most praiseworthy work of God", and others would follow.[149] The significance of this choice would thus have been widely recognised, even if, as Nancy Siraisi points out, Vesalius' account of the muscles of the hand and arm is less prominent in the book and the rhetoric less strident than in Galen's.[150]

This portrait may also represent a direct challenge to Galen. Like him, Vesalius wishes to be seen as more than an expert with the knife, for this is shown on the table alongside pen, ink and paper, the implements of a teacher and author, and, like Galen, he is eager to demonstrate and record the wonders of the divinely created human body. But there is a major difference between the two men. Galen had worked on animals, especially the barbary ape, and had little opportunity to cut into and examine a human body except in his work as a surgeon. In this portrait, the shoulder and arm of the female figure dominate the right hand side of the image, and the way in which the cut fibres and Vesalius' own fingers hang over the edge of the table emphasises that this is a dissection of a human body, not of an animal.[151] Vesalius is at one and the same time acknowledging his great predecessor and announcing that he is doing something that Galen had not himself done. What in Galen was a desirable project that he himself could not carry out is now fulfilled, and his results surpassed by his successor. Taken together, the two initial images of the *Fabrica* throw down a challenge that is continued throughout the book and visibly expressed in its illustrations.

Some of the features of the strictly anatomical images have already been mentioned: their number, their size, and the way in which the varied texture and the three dimensionality of the bones and organs are conveyed by the delicate use of shading. Visibility, legibility and specificity are Vesalius' constant concerns. Vesalius thinks visually. In Book 1, Chapter 4, he insets into the text small figures, to illustrate what Galen meant by the hinge he called '*ginglymus*' and, a little later on, to make clear the different types of suture.[152] John Caius took great delight in pointing out that Vesalius had misunderstood Galen's comparison, and, after it had been corrected in the second edition, he used it to imply a much greater agreement of Vesalius with his own ideas on anatomy than there was in reality.[153] The large size of the page allowed Vesalius to portray whole bodies as well as, on occasion, several individual bones or parts of organs on the same page. So, for example, by viewing different spinal vertebrae together and from different angles or the tibia and femur (Fig. 3.5), the reader could gain an impression of how they were shaped to fit with each other and how the spinal cord and individual nerves pass through them. Similarly, by drawing the skull from different angles and sides, Vesalius provides a clear view of the cranial sutures as well as of the three dimensionality of the skull.[154] The clarity of the image also contributes to the argument. By his imposition of a human skull on top of the lower jaw of a dog (Fig. 3.6), one can immediately see the difference between the two, thus confirming Vesalius' claim that Galen had confused human and animal anatomy in his description of the jaw.[155]

The whole-page images of the human skeleton seen from front, back and side, depend on, but also improve on, those in the *Tabulae Sex*. They are both clearer and more accurate. They still, however, show some distortion in the elongation of the body and the position of the shoulders and skull, probably because the artist was working from a skeleton that had been prepared and then suspended to give an illusion of normality. They are posed artistically in a way that brings out more obviously than before the three different actions that they represent.[156] The first figure (Fig. 3.7) rests his arm on a spade against a background that contrasts bare hills with the woods and verdure of pastoral Italy, a conceit that hints at the gains to be made by hard work. The third figure is hunched up as if in pain and his legs are bent, while the background is ambiguous. The ground is stony, while the trees are misshapen, halfway between life and death. A branch may be growing from a trunk that is partly hollow, and, although some of the sub-branches bear leaves, others are bare. This ambiguity is enhanced by two other details. In the right foreground, blades of grass struggle to emerge from the bare earth, while on the left one can see a lizard, a symbol of both death and resurrection, climbing a rock face. The outcome of a painful illness long remains in the balance.

Still more complex is the middle image, which shows a skeleton with the right hand resting on a skull while the left arm is bent to allow the hand to touch its jaw in a pose of melancholic meditation (Fig. 3.8). It is a thoughtful pose used in portraits of other contemporary humanists, including Durer's famous portrait of Erasmus. The skull, along with the hyoid bone and two tiny bones in the ear, the malleus and incus, all described at length in the book, is placed on top of a large classical tomb in such a way that the reader can see that its top has been cut off and the brain removed. It is an empty shell. In the *Fabrica*, the tomb bears an inscription with a line that was at the time believed to have been written by the Roman poet Virgil: genius lives on: the rest will belong to death. The quotation comes from an elegy on the death of Maecenas, the Roman senator and friend of the Emperor Augustus, who was also the patron of Virgil and Horace.[157] It reinforces the message of the now

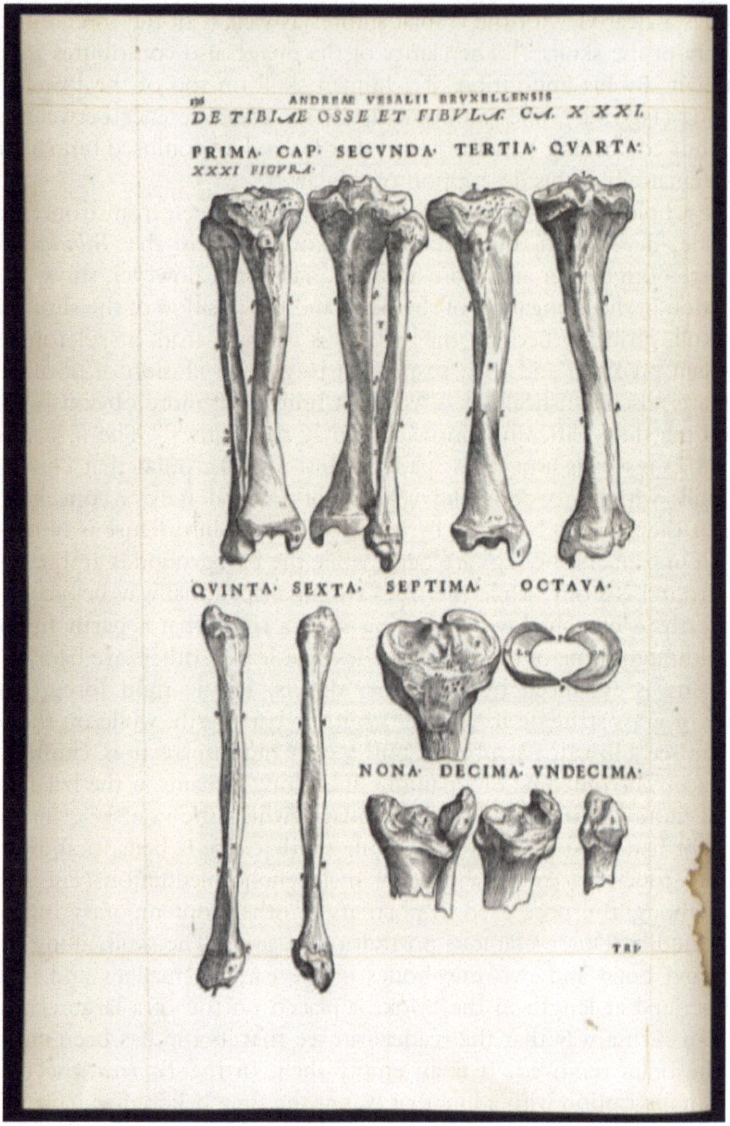

Fig. 3.5 Andreas Vesalius, *De humani corporis fabrica*, Basle: 1543, p. 136. The tibia and fibula from different angles. (*Source* Wellcome Collection)

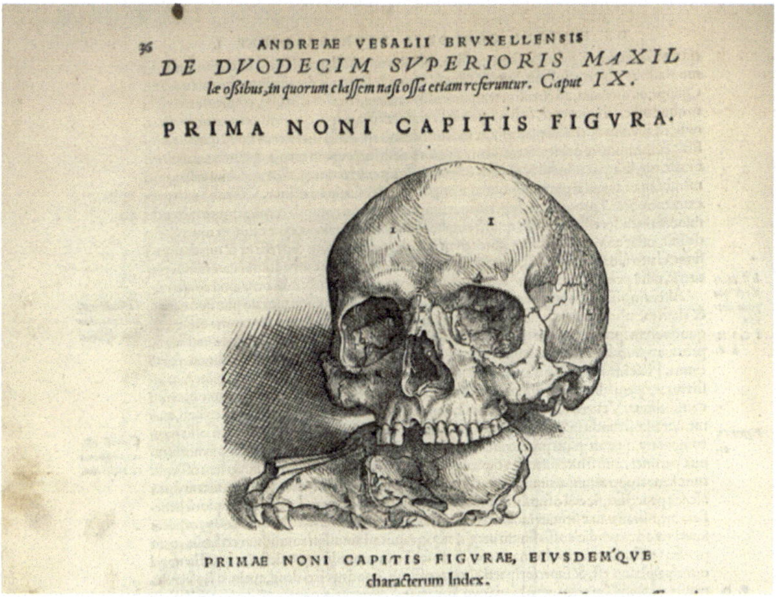

Fig. 3.6 Andreas Vesalius, *De humani corporis fabrica*, Basle: 1543, p. 47. The human skull rests on the jaw of an animal to illustrate the difference between them. (*Source* Wellcome Collection)

empty skull, while at the same time making claims for the author's own genius. The quotation chosen for inclusion in the block in the *Epitome* comes from a slightly later Roman poet, Silius Italicus (26–102), and is a more conventional lament for the passing of beauty at death: "All is dissolved by death, and a Stygian darkness envelops snow-white limbs, and renders vain all homage to their beauty". Why a different quotation was used is unclear, but Vesalius may have simply wished to distinguish the two books further.

Least studied of all the images are the historiated, story-telling initials that open books and chapters in the *Fabrica*.[158] Mediaeval manuscripts, particularly those commissioned by wealthy patrons, regularly began with an illuminated initial and some contained similar illuminated initials marking each subsequent treatise or book. In the most expensive codices, these initials might represent specific incidents or individuals, flowers

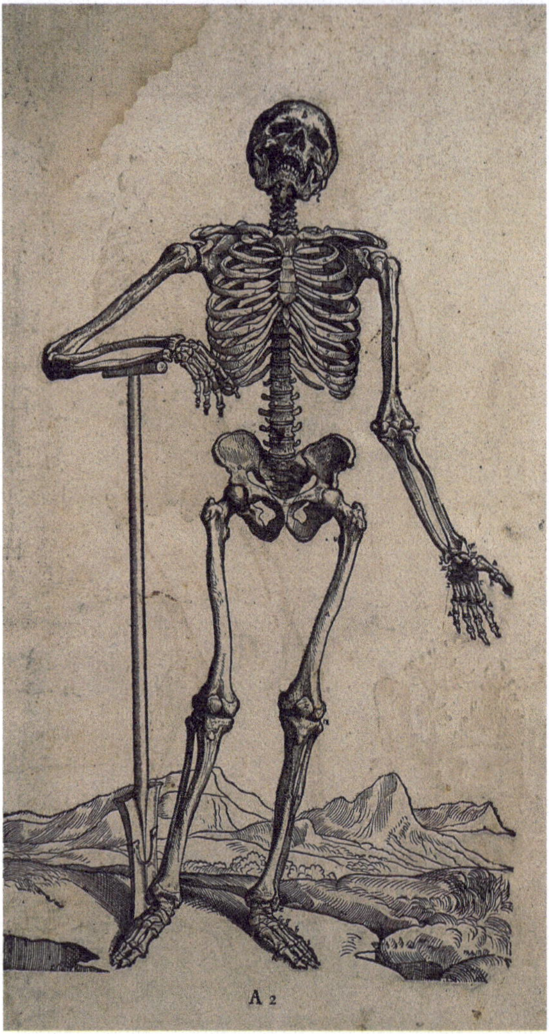

Fig. 3.7 Andreas Vesalius, *De humani corporis fabrica*, Basle: 1543, p. 163. The skeleton from the front, as printed from the original wood blocks by Andreas Maschenbauer, *Andreae Vesalii Bruxellensis, Deß Ersten, Besten Anatomici, Zergliederung Deß Menschlichen Körpers. Auf Mahlerey und Bildhauer-Kunst gericht*, Augsburg, 1706. (*Source* Wellcome Collection)

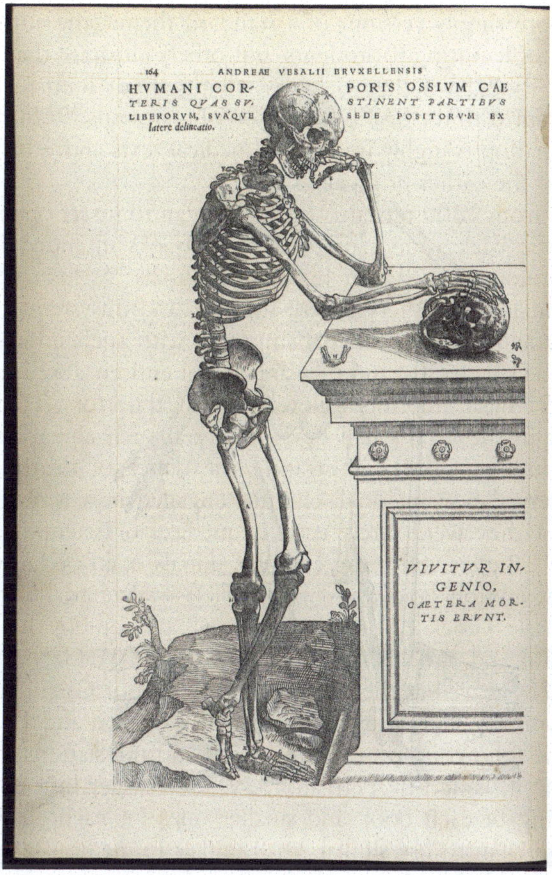

Fig. 3.8 Andreas Vesalius, *De humani corporis fabrica*, Basle: 1543, p. 164. The skeleton from the side. (*Source* Wellcome Collection)

and plants or scenes of daily life. Such extravagance was rare in medical manuscripts, which were mostly produced for use rather than ostentation.[159] One remarkable exception and one of the largest of all illuminated medical manuscripts, Dresden, Sächsische Landesbibliothek Db 92–93, was commissioned around 1470 by Guillaume Poirier, doctor to the King of France. The initial letters of each individual book in this collected edition of Galen are decorated with small vignettes that tell a story. Some are simply versions of a standard theme, the author writing in his library or lecturing to students, but others illustrate the title or first words of the book, whether Galen is talking about foodstuffs or diagnosing a patient with the deadly condition of marasmus.[160] Such a prolific use of illumination is highly unusual in medical texts and was beyond the capabilities of the earlier printers.

Gradually from 1500 onwards printers began to insert opening initials into an artistic background, particularly in books intended for a readership of connoisseurs.[161] But historiated initials, "a luxury" and "the pride of all major publishers", were still rare and unsystematic in 1543. Petrucci Nardelli's list of Venetian imprints with such initials does not begin until the late 1530s, and they are almost entirely absent from scientific works.[162] Basle imprints follow a similar trajectory. Hans Holbein the Younger in the 1520s had devised several such alphabets for Basle printers, including a series illustrating the dance of death and, more relevant, a series showing small cherubs engaged in a variety of playful activities, but they were rarely used in medical or scientific texts until the 1540s.[163] Bigger and more complex initials open prefaces and individual books, editions and translations of classical texts, beginning with the major Basle Greek editions of Aristotle, put out by Johann Bebel in 1531, Plato and Proclus, published by Johann Walder in 1534, and Galen in 1538, as well as the substantial selection of Galenic texts translated into Latin by Cornarius and printed by Froben and Episcopius in 1537.[164] The folio edition of the medical Latin translations of Constantine the African, issued by Petri in 1539, used larger historiated initials at the opening of each book and smaller ones for each chapter. A few printers also began to use similar small initials in medical texts. Isingrin, for instance, in 1535 had used small initials featuring one or two figures at the opening of each *Letter* in his edition of Manardo's *Epistolae Medicinales*, *Medical Letters.*[165] The following year Lasius and Thomas Platter did the same for the opening initial in each book of Guinter's *Principles of Anatomy.*[166] By the time that Oporinus became more involved with the firm, such initials were common, at least for the opening of individual books, as in Oporinus' 1541 printing of the Greek Theophrastus and Robert Winter's 1542 publication of the fever tracts of Alexander of Aphrodisias and John of Damascus.[167] In 1542, Oporinus used historiated initials for the first letters of the introductory treatises and the first book of the translated text in his controversial printing of the Koran in Latin, showing *putti* fighting (M) jousting (I) and bringing home a captive putto tied to a pole (Q).[168] Shortly after setting up the Koran, in March 1542, Oporinus used three other historiated initials with putti (D) and a large and a small M, to open the preface and the two books of a treatise by a Paduan colleague of Vesalius, Bassiano Landi.[169] Landi's anatomical treatise *De humana historia, uel singularum hominis partium cognitione, An Enquiry into Mankind, or an Exposition of the Individual Parts of Man,* is a long book in a small format, expounding the structures and functions of the body from a Galenist perspective.[170] The text itself is unillustrated, although, in a passage that hints at the progress of the future *Fabrica,* Landi accepted that at times a picture might be more effective in describing the muscles than any words.[171]

By 1542, then, initials wittily showing one or two figures, often cherubs, acting in various ways, were being used by Basle printers, although rarely for books by contemporary authors. Oporinus himself had several different sets of such initials to hand.[172] But the decision to introduce a whole programme of such initials into the 1543 *Fabrica,* heading chapters as well as individual books, goes far beyond what Oporinus had done before and must have been the result of further discussion between author and publisher. Vesalius' initials are an integral part of the book, not scattered about or confined to preliminary matter, and their subject is clearly related to the whole theme of the *Fabrica.* They are not stock images, unlike the much smaller initial headings in the index, which have nothing to do with anatomy, or those used to begin sections in Colombo's 1559 anatomy book, but they were freshly created as part of the project to introduce anatomy to a wider audience. In this, as John Goree pointed out, there is another link with Basle and with Isingrin's publication of Fuchs' *De historia stirpium,* a work that would have been well-known to Oporinus.[173] It too has historiated initials for the opening letters of many of his chapters and sections that show putti engaged with plants, but, although of a similar size, these are more crudely designed. It is not difficult to see here Oporinus and Vesalius taking their inspiration from the earlier book and setting out to surpass it artistically and typographically.

If, as has been suggested, one of the artists working with Vesalius in 1541–1542 was Giuseppe Salviati, Vesalius would have found in him a young artist who had already experienced the constraints of having to design his images to fit into a varied space. Some of those he had designed for the fortune-telling cards depicted in Marcolini's *Le Sorti* consist of only a single figure, but others are more crowded.[174] Vesalius' images are even more complex, often involving several figures of cherubic *putti,* and tell a story at the same time. Like the larger images within the book itself, they will have been first drawn and then copied before being placed on the wood block on which they would be engraved once the size of the space was known. Their tiny details would have demanded a high degree of expertise from both artist and engraver.

There are altogether 21 such initials, representing all the letters in the standard Roman alphabet except for B, X and Z. Four of them, I, O, Q and T, were also cut in a larger size to distinguish the opening initial in each book. Once cut, the initials were reused for each subsequent repetition of the individual letter, with one exception. At p. 228/328 the initial L (Fig. 3.9) derives from the alphabet series with putti designed by Hans Holbein the Younger and familiar to Basle printers from 1524 onwards, including Cratander, whose working stock had been bought up on his death by Platter's consortium.[175] Oporinus had already used two of the initials from the same set for the I and Q in the prefatory matter to his edition of the Koran in Latin. Here it replaces that used earlier for L, probably because its subject, a cherub defecating, is very appropriate to this chapter, III, 51, which deals with the muscles of the rectum. It is also slightly larger than the other images and appears a somewhat tighter fit, which suggests that it was inserted as an afterthought to replace one of the original wood blocks after the text itself had been set up. Since the original initial (Fig. 3.10) continued to be used for later repetitions of the letter L in the book, the introduction of the new letter must have been a deliberate decision, or at least met with acquiescence, on the part of Oporinus or Vesalius, for it was the only such initial to be retained in the second edition of 1555.[176] The much smaller headings in the index (Fig. 3.11), apart from the opening initial A, have nothing to do with anatomy but will have formed part of the printer's collection of stock initials.

These images wittily restate the author's professional activities as an anatomist and surgeon by transferring them to an imaginary classical world where *putti* set broken bones, perform a catheterisation, cauterise a patient's head and carry out dissections.[177] They disinter a corpse and bring it away on their shoulders, and surreptitiously gather up bones in a cemetery, later carefully boiling a skull to remove any flesh and hair. They vivisect a pig, as Galen had done and Vesalius had depicted in one of his last images. They catch an animal, probably a dog, and hang it up before dissecting it; they open a human thorax, remove a foetus from a

DE RECTI INTESTINI MVSCVLIS.
Caput LI.
RECTI inteſtini portio cũ ſuis muſculis à corpore reſecta, quã dũ reliqua eximimus inteſtina, in corpore adhuc relinquere ſolemus.
A Rectum inteſtinum, quod in libri quinti tabulis euadet conſpicuum, præſertim autem octaua figura & uigeſimaſecunda.
B, C Duo muſculi, inteſtinum poſt egeſtionem ſurſum trahentes.
D Muſculoſa ſubſtantia penis radici adnata, in mulieribus porrò ad infimam pudendi partem commiſſa.
Inſuper D ſedes notatur, à qua in penem muſculi duo pronaſcũtur.
E Circularis muſculus, ne præter arbitrium excrementa deijciantur præcauens.
LAVDATA nunquã ſatis Natura recti inteſtini extremo muſculos largitur, qui in fecium excretione protruſum recti inteſtini finem, alacriter ſurſum reuellunt: dein alium dat muſculũ, qui intempeſtiuam inuolun tariamq; excrementorum egeſtionem prohibet. Muſculi itaq; ſedem ſurſum attrahentes, lati & tenues, neq; admodum carnoſi autrubentes ſunt, uix ab aliquibus oſſibus originem ducentes. A ligamẽtis enim ex pubis oſſibus, & coxendicis oſſe, & ſacro exortis, quæ huius regionis muſculis obducuntur, ipſaq; demum peritonæi infima ſede utrinque ſinguli enaſcuntur muſculi, qui deorſum protenſi, rectum inteſtinum amplexantur, exti mæ ipſius tunicæ inſerti. Præſentes muſculi quum lati admodum ſint, & ſuis finibus totam inteſtini latitudinem ambiant, ipſorum portio radici quoq; penis, mulieribus uerò uteri collo adnaſcitur. Quod nonnulli perpendentes, tres ſedem ſurſum agentes muſculos recenſuerunt. Vtcunq; uerò ſit, omnes horum muſculorum partes pari præficiuntur muneri, quo ſedẽ

Fig. 3.9 Andreas Vesalius, *De humani corporis fabrica*, Basle: 1543, p. 228/328. This section on the muscles of the rectum is introduced by a unique initial L showing a defecating *putto*. It comes from a set of letters that was originally created for another Basle publisher by Hans Holbein the Younger. (*Source* Wellcome Collection)

LARYNGIS
ſum, ſed uarijs m
nonnulli à literæ
qua uoce diſſecti
terunt. Alijs rurſ
etiam nomine int
uertere conſueuer
ex Galeno, quem & Italia & Germania l

Fig. 3.10 Andreas Vesalius, *De humani corporis fabrica*, Basle: 1543, p. 55. The initial L as usually used shows the removal of the body of an executed criminal from the gallows for transport to a dissection. The hooded figure holding a cross indicates that this is an officially sanctioned removal and that the remains will be given a Christian burial after the dissection. (*Source* Wellcome Collection)

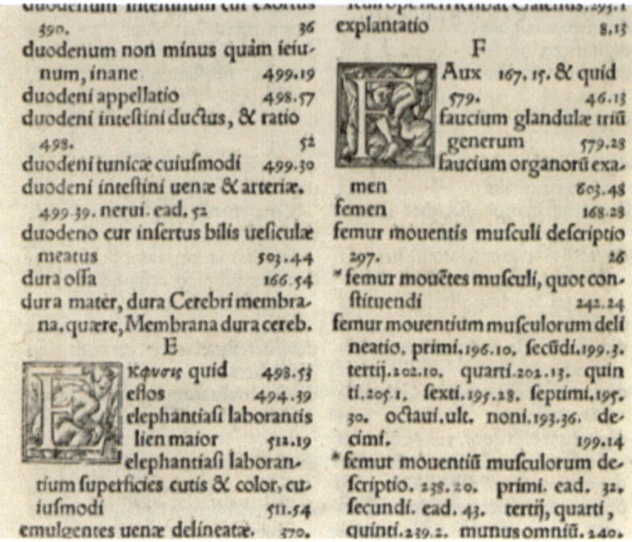
390. 36
duodenum non minus quàm ieiunum, inane 499.19
duodeni appellatio 498.57
duodeni intestini ductus, & ratio 498. 52
duodeni tunicæ cuiusmodi 499.30
duodeni intestini uenæ & arteriæ. 499.39. nerui. ead. 52
duodeno cur insertus bilis uesiculæ meatus 503.44
dura ossa 166.54
dura mater, dura Cerebri membrana, quære, Membrana dura cereb.
E
Κφυσις quid 498.53
eflos 494.39
elephantiasi laborantis lien maior 511.19
elephantiasi laborantium superficies cutis & color, cuiusmodi 511.54
emulgentes uenæ delineatæ. 570.
explantatio 8.13
F
FAux 167.15. & quid 579. 46.13
faucium glandulæ triũ generum 579.28
faucium organorũ examen 603.48
femen 168.28
femur mouentis musculi descriptio 297. 26
* femur mouẽtes musculi, quot constituendi 242.24
femur mouentium musculorum delineatio. primi. 196.10. secũdi. 199.3. tertij. 202.10. quarti. 202.13. quinti. 205.1. sexti. 195.28. septimi. 195.30. octaui. ult. noni. 193.36. decimi. 199.14
* femur mouentiũ musculorum descriptio. 238.20. primi. ead. 32. secundi. ead. 43. tertij, quarti, quinti. 239.2. munus omniũ. 240.

Fig. 3.11 Andreas Vesalius, *De humani corporis fabrica*, Basle: 1543, f. Ll 3v. Two small initials from the Index. (*Source* Wellcome Collection)

dead bitch and inflate a lung just as Vesalius himself advocated. Sometimes they examine a body (or view the result of an enema), while other *putti* look carefully at a book.[178] The large opening initial Q of the dedication to the Emperor almost parodies the dissection scene on the title page (Fig. 3.12). A crowded audience of naked *putti* gather to watch the dissection of a pig, strapped to a table as recommended in Book VII, 2, while another *putto,* like the student on the title page, stands by with a book to compare it with what he is seeing. Six others, like angelic supporters in royal and religious art, gaze attentively at the gruesome spectacle about to take place.[179] The two angels in the outer top corners are balanced by playing *putti* in the two lower ones. This is a witty counterpoint to the serious description in the book. If, as Bette Talvacchia has argued, readers of Estienne's anatomy books would be expected to recognise the pornographic images and texts that literally encompass the drawings from his dissections, Vesalius' readers would also be expected to appreciate the joke of having the serious work of surgery and anatomy carried out by over-large toddlers.[180]

Fig. 3.12 Andreas Vesalius, *De humani corporis fabrica*, Basle: 1543, f. 1r. This large initial Q, the first in the book, wittily reinterprets the dissection scene on the title page, above, Fig. 3.11. (*Source* Wellcome Collection)

But problems still remain over the nature of Vesalius' involvement in these small pictures. Some incidents clearly record his own experiences or advice given in the *Fabrica,* but others, like the surgical scenes, may have been the artist's own suggestion. Most striking of all, very few of the almost 200 images in the initials relate to the chapters that they head. This is not surprising, for once a letter had been cut in a block it would have been reused for the same initial, no matter what the context of the chapter. But few of the images that clearly relate to a specific chapter or comment appear at its opening, even though it would have been possible for Vesalius to have indicated this to the artist from the start or modified his wording. It is only with the last two initials in the first book, C, carrying a corpse to being stripped of flesh in a stream, and P, collecting bones in a cemetery, that what is depicted in an initial in that book becomes at all relevant to the subject of the chapter.[181] Even if an image fits the subject of the chapter, it is not certain that it was originally designed to go there. The letter D is used twelve times as an opening

QVONIAM Galenus in horum neruorum descriptione, quemadmodum & nonnunquam alibi, mentionem facit instrumenti glossicomi, nos lectori non incommodaturos duximus, si hunc characterem F, in quo id instrumentum obiter exprimitur, hic interijceremus: una cum E, quo instrumentum delineatur brachij, sed potissimum femoris luxationibus restituendis idoneum. Licet hæc alio libro ac argumento tractanda putemus.

Reliqua sexti paris in thoracis amplitudine series.

...us obuius, ipsi arteriali uenæ adnascitur, ... magis exporrectius in cordis basim fertur.

Fig. 3.13 Andreas Vesalius, *De humani corporis fabrica*, Basle: 1543, p. 329/429. In the margin Vesalius in Padua instructs Oporinus to have two new initial letters cut to indicate how a box-splint should be used. (*Source* Wellcome Collection)

initial. In a crowded scene around the letter medical *putti* use a saw to cut away the bone at the top of a head in preparation for dissection of the brain. But it is only with Chapter 14 of Book III, a discussion of the blood vessels in the brain, that the design links with the text, and only at the beginning of Book VII, Chapter 4, can it be said to portray directly what is described verbally, the opening of the skull. While all the themes of the initials are relevant to something mentioned in the *Fabrica*, there is little indication that Vesalius had attempted to link them with any of the chapters they headed. Their overall message mattered more than their relevance to a specific chapter.

There are three or four other exceptions to this lack of fit. One, the Holbein initial, has already been discussed.[182] The second, the initial I that opens Book III, Chapter 11, p. 294/394, may have been drawn originally for it, as Faller argues.[183] It shows cherubs removing the placenta from a parturient woman to place it in a large bowl. The chapter itself does not mention the placenta, although its subject, the umbilical vein, is certainly relevant to the image. The two other examples occur together on p. 329/429, near the beginning of II, 51, in a passage sent as an afterthought to Oporinus and inserted at the side of the text (Fig. 3.13). In it, in order to make clear to his readers just what Galen had meant by this technical term, Vesalius demanded that his discussion of the Galenic word *glossokomion*, 'box-splint', be illustrated with images of this support being used in practice. As a result, all other instances of the initials E and F show it being used to deal with broken or dislocated bones, although such operations play only a minimal role within the *Fabrica* and Vesalius reserved a more detailed discussion of them for a later treatise, probably on surgery, that was never completed. This seems to have been a hurried decision on the part of Vesalius, for although the word appears in Galen's commentary on this surgical procedure, it does not appear in the text of the *Fabrica* itself. Not surprisingly, the passage in the margin was removed from the second edition, although the subject continued to be used in the recut letters E and F. Vesalius' instructions from afar must have reached Oporinus before the original blocks for these letters had been cut and used in the printing of Book I, where they appear first on pp. 45 and 78, respectively. This confirms the suggestion of Willy Wiegand that the designs for the opening initials were, at least in part, not sent with the larger wood blocks from Venice in the summer of 1542 but were instead passed on as drawings to be turned into blocks in Basle.[184] Once cut, some of the initials were reused by Oporinus in other similar publications, such as Bassiano Landi's tract on medicine later in 1543 and Vesalius' defence of the *Fabrica* in his letter on the China Root.[185]

The anatomical images in the *Fabrica* are deservedly praised as works of art in themselves, and the historiated initials deserve similar approbation for the skill of their artist in massing so many figures within a confined space, something visible also in the scene on the title page. The enlarged modern reproductions of the initials in the recent English translations have emphasised still more their wit, their artistic quality and their singularity within a medical book. As a way of depicting the human body Vesalius and his artists fundamentally changed notions of what was possible. But, at the same time, they conveyed more than anatomical information linked closely to a detailed text. They sent out a message about the author as a man of culture, learning and taste, the successor to Galen, if not his supplanter. Both the title page and the historiated initials revealed anatomy, the understanding of the human body, as something of interest for everyone, and certainly for men who could afford to buy the *Fabrica* and appreciate its artistry as well as learn from the images that they could see.

The Message of the *Fabrica*: The Text

Vesalius' choice of Latin for his treatise was not surprising. It was the universal language of the European cultural and medical elite, a social as well as an intellectual marker of high status. Although the rapid development around Europe in the sixteenth century of town grammar schools or gymnasia that taught Latin meant that more people were now able to read books written in that language, there was still a barrier between the Latinate physicians and many of the surgeons and barber-surgeons who were literate only in their own local language. Vesalius was aware of this division, ensuring a German version of the *Epitome* of the *Fabrica*, but a complete version of the great book in a modern vernacular had to wait for almost half a millennium.[186]

Vesalius had been trained at some of the best schools in Northern Europe since childhood to write and speak the new humanist Latin, and his fluency in it is obvious.[187] Yet it also presented both a challenge to him and a barrier to others. His Latin is not always easy to follow. Its long periodic sentences, often taking up several lines in an unbroken block of printed prose, have a complex structure, which, together with the need to refer frequently to notes on the illustrations, makes reading deliberately slow. Vesalius uses a wide vocabulary that often alludes to passages from Roman authors to give added weight. He frequently couples two or three words together to emphasise a point and uses a wide range of synonyms and grammatical structures to show his skill in *variatio* (a key Renaissance index of style) and to avoid wearying the reader with endless repetitions of the same word. Like Galen, he adopts a variety of strategies in his discussions. Will Richardson singles out for approval the eloquent praise of the Creator's design of the thorax, which calls to mind the 'Hymn' that closes the last book of Galen's *On the Usefulness of Parts*.[188] The opening page of his dedication to Charles V is the equal of anything in Cicero in its rhetorical power allied to a real solemnity and a choice of words to match.[189] The whole book is a literary as well as a medical *tour de force*.

At the same time, Vesalius was having to create, or at least use, a technical vocabulary that had only just been (re-)introduced and that was still fluid. His Paduan colleague Landi in 1542 declared that all available translations from the Greek were often weak and misleading, and that he was forced to employ his own new terminology, for which he expected to be criticised.[190] As we have already seen, in the verbal descriptions of the *Tabulae Sex* Vesalius still thought it necessary to indicate the same technical term in different formulations, but by 1542 he was much more consistent in his terminology. A love of variation may explain the variety of words used to indicate the type of branch of the blood vessels, 'small branches', 'twigs' and 'twiglets', but he often keeps the same name for individual parts of the body.[191] His oscillation between '*vena brachialis*' and '*vena humeraria*' for the vein in the upper arm, the former more Greek than the latter, is unusual, and he avoids the traditional Arabist terms.[192] This consolidation of a humanist vocabulary is in part due to the wider acceptance of these terms than had been so a decade earlier and in part to his own greater familiarity with the new Galenic translations through his involvement in the Juntine edition of Galen. It was one of the achievements of the *Fabrica* to have consolidated still further this relatively novel terminology.

Any account and assessment of what Vesalius wrote is hampered by two related difficulties: his silence and our own ignorance of what was being discussed or taking place in the years immediately before 1543. Levi Lind's translations and studies of 'pre-Vesalian' anatomical writings remain fundamental to any investigation into his sources, but they deal only with printed material from a handful of universities.[193] The student notes on Corti's lectures at Bologna in 1540, for which Vesalius acted as dissector, may provide a better indication of what was typical, while Guinter's own 1539 revision of his *Principles of Anatomy* shows how much was changing even as Vesalius was writing.[194] Giovanni Battista Canani's study of the muscles of the hand and arm is a portion of a (never completed?) larger work that raises almost as many questions as it solves about anatomy teaching at one university and in one city, Ferrara, in the early 1540s.[195]

The 1542 treatise *De humana natura, On Human Nature,* by Vesalius' Paduan colleague Bassiano Landi, while not disdaining the results of dissection, is a philosophical survey of the parts of the body, with particular reference to their function and utility in the great scheme of the divine Creator.[196] Its novelty consists in a new and more accurate Latin terminology based on a more sophisticated knowledge of Greek, but in its (to modern eyes old-fashioned) exposition of the results of dissection, it may, like Corti's lectures, be more typical of what was being taught in formal anatomies than the *Fabrica.* But beyond Paris and the Po valley our ignorance is near total, even about universities such as Leipzig, Marburg and Wittenberg where we know that anatomical teaching was taking place. The gap between printed books, like those of Estienne or Laguna, and their actual lectures may well be wide.[197]

Vesalius' silences are more difficult to penetrate. Like Galen, he is a master at dissimulation, deliberately disguising his debts to others.[198] In his preface he accepts that things have somewhat improved in the previous few year, but he gives no details of the individuals or institutions involved, thereby emphasising his own role in this development.[199] Alongside a few friends and students who brought him interesting specimens or made favourable comments, and by definition were not experts, only two already published authors are mentioned in the text by name.[200] Balamio, the editor and translator of Galen's tract *On Bones,* is stigmatised for his refusal, and that of his patron, Cardinal Ridolfi, to let Vesalius see the manuscript of Galen's *On Bones* in their possession. The other, the Marburg professor Joachim Dryander, is mentioned only for consulting Vesalius about a problem he had encountered in comparing a passage in Hippocrates with his own observation of the omentum.[201] While acknowledging an epistolary friendship, Vesalius introduces their correspondence only to emphasise his own different and more correct view. His successor at Padua, Realdo Colombo, is praised for bringing to his attention further examples of anomalies in the hyoid bone, but in a way that makes clear the priority of the master over his pupil.[202]

In the text itself there is no word of such writers as Benedetti, Berengario, Massa or his Parisian teachers, and Corti may not have been the only nameless physician "born to decorate a professorial chair", who is accused of failing to benefit from any attendance at a Vesalian demonstration.[203] Unlike Galen, whose arguments with Lycus, Julian and sundry philosophers and doctors often allow us to reconstruct their ideas, Vesalius lumps his opponents together anonymously to be castigated for their errors or their reactions to Vesalius' earlier work. They are "windbags" and "idiots", for these fierce and voluble upholders of Galen lose their tongues when faced with the evidence of a corpse. They sit aloft, like Prometheus, pronouncing from their high chair their imaginary descriptions of the body while turning their backs on the corpse in front of them.[204] If they do carry out a dissection, their slapdash procedures only serve to confirm Galenic error.[205] These "stinkers" argue about trivialities, rejecting the evidence of dissection as irrelevant to the needs of doctors and jealously concocting plots against Vesalian interpretations out of sheer hatred.[206] Building on Galen's belief that there was a bone in the centre of the heart, they pile fantasy on fantasy.[207] They misunderstand what Galen actually said when he was correct, and, even when they themselves are right, Vesalius can find an excuse to condemn them. Disputes among them about the organ of smell are the result of a lack of dissecting experience.[208] Italian university anatomists who criticise earlier writers for confusing the gloutia of the brain with the vermis have trumpeted this as a major discovery, even though it had already been carefully described by Galen and is a trivial correction. He offers a simple excuse for giving no names: by not taking issue with his contemporaries directly, Vesalius can, be insists, avoid a reputation for contumacy or prolixity.[209] But, unlike Galen, he does not allow us to judge their objections or, still more so, his dependence on them, thereby creating, as his strongest modern critic, Juan Barcia Goyanes, put it, the "myth" of Vesalius.[210]

Vesalius, Aristotle and Galen

His view of the history of anatomy is largely taken over by Galen, often *verbatim*. Although he claims to have gained from reading his grandfather's notes on Hippocrates, his praise of Hippocrates as the father of anatomy and his citations of him are entirely Galenic.[211] Indeed, far from quoting Hippocrates directly at times, he relies on passages cobbled implausibly together by Galen from different treatises.[212] His attitude to his great predecessor Aristotle is similar. Galen agreed that he was a remarkable anatomist who carried out some excellent experiments in dissecting a wide range of animals, but he was not working with humans, something reflected in his many errors.[213] Nonetheless, his emphasis on function and purpose, his teleology, was undoubtedly correct, and he laid the foundations for the work of his Alexandrian successors, Herophilus and Erasistratus and the like. Without Aristotle, the pursuit of anatomy would have scarcely begun. Indeed, in a prophecy of what his own successors such as Fabricius of Aquapendente were to do later, Vesalius predicted that by following Aristotle's example future students would compare the bones of humans, animals and birds to discover how they walked or stood.[214]

His attitude to Galen is more complex and more ambiguous. On the one hand, he is keen to express his own credentials as a Galenist, not least because of his work on editing and improving the Latin versions of his anatomical writings that had been so badly published by Italians and Germans.[215] Some of his notes on Galen he planned to publish after finishing the *Fabrica,* but these, like much else, were put aside and probably destroyed when he entered imperial service.[216] He is proud to call himself a Galenist and trusts that all true doctors will follow him in acknowledging this physician with a passion for dissection who based everything he wrote on anatomical foundations.[217] He can lavish praise on Galen. He is the author of all good things, the prince of anatomists and the chief among the teachers of anatomy.[218] He made valuable observations on the lobe of the lung in animals or the popliteus muscle "hiding behind the knee", and he was frequently correct in describing what he saw.[219] But these compliments are far outweighed by the criticisms, sometimes direct, but often slyly oblique or intended to underline Galen's weakness. The lobe of the lung is found only in animals and is there perfectly obvious, while Galen's claim to be the discoverer of the popliteus muscle is much exaggerated, for the muscle is by no means as important as he believed. Galen frequently challenged the ideas of Herophilus and Erasistratus, who had dissected humans, on the basis of his own animal dissections, but when he did accept their views, it was because that what they had found was obvious even to the beginner. Galen regularly praised the purposeful hand of the Creator, a theme that is taken over frequently and at length by Vesalius, but his choice of what to praise is very often wrong. One of the most celebrated of Galen's paeans to the Creator, the last book of *On the Usefulness of Parts,* is dismissed with the comment that Galen's encomia relate to function, when he should have been explaining the action of each part of the body. Vesalius' own debts to Galen's functional anatomy, especially in the final books of the *Fabrica,* are carefully hidden.[220] Even when Galen acknowledges that the wisdom of the Creator is wonderfully clever, he still can be criticised. The path of the carotid arteries is far more impressive than what Galen had believed from his animal dissections.[221] Indeed, Galen is being impious in his attitude towards Nature by expecting the reader to follow his own convoluted explanation for the movements of the head: Vesalius, of course, will then explain it more succinctly and at greater depth.[222] Everyone should know their Galen, but it is Vesalius who has the greater understanding of the workmanship of the Creator in designing the human body.[223]

Disagreements with Galen are often patent. Galen frequently changes his mind, he is inconsistent, he mixes up animal and human anatomy and his prolixity damages even the best of his arguments.[224] But, more often, Vesalius hides behind rhetorical strategies that would have earned the admiration of a Cicero.[225] He will not accuse Galen of inventing a separate bone between the clavicle and scapula, for that would be impious: it would be better just to say that he misrepresented a cartilage as a bone. (i.e. the error still remains).[226] He accepts, on Galen's authority, that there are three tendons in the seventeenth muscle of the hand, while making it clear that this is something about which he has very strong doubts.[227] Throughout he continues to express his devotion to Galen and emphasises his desire to avoid being convicted of impiety towards him should he be found to have criticised him unduly.[228] But this loyalty has its price, for it has misled him in the past and now still prevents him from pointing out yet more Galenic errors, along with his fear of being condemned as equally prolix.[229] In fact, it is Galen who is responsible for the length to which he is forced to go to reveal his mistakes, for without Galenic interruptions his text would have been much shorter and easier to follow. Besides, had he not argued his case at length, everyone would have thought him a liar for disagreeing with Galen.[230] Even though at times he blames Galen's successors for the disappearance of this earlier human anatomy, Galen is not excused, for it is because others followed his pronouncements and did not attempt to go beyond them that the proper anatomy of the human body was forgotten.[231] It may be true that, as his defender Donald O'Malley pointed out, Vesalius is only correcting the mistakes of his great predecessor, but the lengths to which he goes in asserting his own superiority cast doubt on the extent of his respect for Galen.[232]

The overall message of the *Fabrica* is clear and cogent. Only through dissection, and in particular human dissection, can one gain an accurate knowledge of the human body. It is not enough to rely on a book. 'Navigating by a book' is a phrase often used by Galen in condemning other doctors and writers on anatomy, and its frequent use in the *Fabrica* to stigmatise Galenists is doubly hurtful for that reason. One needs to see and touch a body, whether that is of a human or an animal, to be able to understand how it is constructed and functions. Attendance at a public anatomy is valuable, and, still more, participation in a private anatomy where it is possible to draw closer to the corpse and perceive small details. But best of all is one's own experimentation with the knife, a constant slogan of Galen's, and one that had been largely obscured before the return of the Greek Galen to the West in the last years of the fifteenth century. Even a minor acquaintance with the results of dissection will reveal things about human or animal anatomy that is contrary to what is generally believed and that will confirm the proper anatomical basis of medicine. Whether it is the shape of a bone, the passage of veins or nerves, the location of a specific part or the presence of anomalies, all can best be grasped through the art of dissection.[233]

But it is human dissection above all that should be the guide to understanding the human body. Galen had dissected mainly pigs, dogs and apes, and although his descriptions were generally accurate, apart from a tendency to conflate what he had seen in different animals, the gap between this and human anatomy was substantial. Chapter after chapter lists mistakes by Galen in transferring what he knew from animals to humans. If only, like the Latin writer of an *Enclopaedia of Arts* a century before him, Cornelius Celsus, he had taken proper notice of what the Alexandrian authorities had written, he would have avoided many mistakes in his account of the bones.[234] Indeed, Vesalius suggests that the evidence of Celsus, together with the names given in Greek literature and early medicine, indicates an ancient interest in human anatomy that may precede even the Alexandrians.[235] Galen may have wanted to do more than rely on chance observation, but Vesalius calculated that he had only twice observed a human corpse, a calculation that so annoyed John Caius that he compiled a list of passages on the front of his copy of his own *Libri aliquot graeci*, now in the Cambridge University Library, classmark Adv.d.3.1, to show that Galen had done this far more often than Vesalius had supposed.[236]

The message of the need to dissect humans is amplified further by the image on his book's title page and by the record of his own experiences. Whether he is carrying out a formal dissection or assisting at a private showing, performing an autopsy on a Venetian gondolier or a young girl or grubbing around in the Cemetery of the Innocents in Paris for bones, Vesalius was investigating human bodies.[237] Only at the end of the book do we find any substantial acknowledgement of the need to investigate the body also by using animals. An image showing a pig strapped down ready to be dissected adorns the final chapter which sets out how best to perform a dissection, but there is no hint that, as in Vesalius' Bologna dissections in 1540, far more dogs than humans were being dissected.[238] This was not just because it was hard to find humans to dissect, as Andrea Carlino has shown, but there were also social constraints on who might be subjected to the indignity of a public anatomy.[239] Besides, how the body worked and how individual organs and its systems functioned could only be demonstrated on living things, i.e. animals. Vesalius himself had to rely for his bodies in formal dissections on the goodwill of the authorities who would allow an executed criminal to be handed over for a university demonstration within a short time of his death.[240] This would allow the anatomist to dissect while the viscera were still substantially intact and see traces of bodily activity that would have disappeared soon after. But as in his Bologna demonstrations, animals were essential, and long continued to be so, although later anatomists were more careful to note the differences between humans and animals.

For Vesalius, then, the mistakes in Galenic and, still more, post-Galenic anatomy were systematic. It was not enough, as Berengario or Guinter had done, to point out and correct an individual error; the fault lay in assuming that what could be seen in animals was more or less identical in humans. The proof of this lay in the abundance of new discoveries that he had made, many of them at once obvious to the unprejudiced eye. (His earliest critic, Niccolò Massa, may not have been unfair in saying that many of them had already been made by others, like himself, who had not yet, for often good reasons, succeeded in publishing them in print.[241] But, even with these caveats, the list of his corrections is impressive.[242]

The *Fabrica* opens with the bones, a reversion to Galen's order of dissection against that of Mondino and his followers. Vesalius rejects Galen's description of the upper and lower jaw, showing that he is describing the anatomy of the dog, as well as his ideas of a seven-boned sternum and a bone in the heart.[243] His description of the tiny sesamoid bones in the wrist is substantially complete, but others before him may well have disagreed with Galen's opinion that the patella was partly cartilaginous.[244] He is equally critical of Galen's description of the muscles of the jaw and takes pains to point out where Galen had been describing animal rather than human muscles.[245] His account of the vascular system, however, in Book III is still heavily influenced by Galen, even if it is more accurate, because, like all anatomists before Harvey, he viewed the arterial and venous systems as independent. But he argued that the vena cava took its origin from the heart, not the liver, because the size of the vena cava was greater there, although he accepted that it acquired blood from the liver and regarded any dispute about what he described as an offhand remark as trivial compared with his more radical conclusions about the bones and muscles.[246] The later volumes, which deal more with function than the earlier ones, are much more Galenic but also reveal more of Vesalius' hesitations about interpreting some of the differences he found. While in general his description of the nervous system is superior to that of Galen, he still puzzles about what, at this stage, he still regards as anomalies, for instance, in the optic chiasmus, or the result of illness or as in some way necessary to explain the workings of the body.[247] He rejected the existence of the rete mirabile, something he had accepted as late as 1540, but he still sought to locate the area in the brain where the vital spirit from the arteries was turned into animal spirit.[248] But he was hampered throughout by an inability to dissect enough human bodies, especially female ones, and to do so at a time when they still retained enough information to show how they had lived and worked.

For all its faults, the *Fabrica* is a brilliant production, far in advance of any of its competitors. To say this is not to indulge in Vesaliolatry, to use a phrase of Roger French, and to consider it out of its intellectual and medical context. Rather, it reflects the view of many of his contemporaries as well as that of most modern scholars, whether they have been concerned with the history of medicine or the history of the book. Whether one chooses to focus on Vesalius' anatomical discoveries and his overall message, his images or the complex creation of the book itself as a printed object, everywhere one can find Vesalius doing something that is unusual and setting standards that few of his contemporaries could emulate. It is their extent and, even more, their combination that justifies the accolades given to him and his book.

Notes

1. Rinaldo F. Canalis, "Vesalius' Methods in the Production of the *Fabrica* with Emphasis on the Neuroanatomy Chapters," in *Andreas Vesalius and the "Fabrica" in the Age of Printing: Art, Anatomy, and Printing in the Italian Renaissance*, edited by Rinaldo Canalis and Massimo Ciavolella (Turnhout: Brepols, 2018), 131–70, rightly draws attention to the speed with which the book was written.
2. Andreas Vesalius, *De humani corporis fabrica* (Basle: J. Oporinus, 1543), vi. For later additions and the initials, see below, pp. 99, 101.
3. Finding the exact number of words is difficult, given that some pages are given over entirely or largely to images. I have taken 500 pages of 57 lines as the base for my calculations, and an average of 12 words per line (the range on the pages I have checked is between 8 and 15). I have not considered the index, but this figure, almost certainly an underestimate, gives some idea for comparison with other books. O'Malley, *Andreas Vesalius of Brussels, (1514–1564)* (Berkeley and London: University of California Press, 1964), 135, suggests around 5000 characters and approximately 5 hours per page for setting.
4. Vesalius, *Fabrica,* vii: Roger Gaskell, "The Typography and Layout of Vesalius' De Fabrica as Specified in his Letter to Oporinus," *L'Illustrazione, Rivista del libro a stampa illustrato* 3 (2019), 42–43.
5. Niccolò Massa, *Epistolae medicinales et philosophicae* (Venice: F. Bindoni and M. Pasini, 1550), 52v, 58v.
6. Canalis, "Methods", 137. He does not include the folded page 313/314 which could be removed and turned into a flap anatomy in the same way as an anatomical fugitive sheet. Not all surviving copies contain this sheet.
7. John B. de C. M. Saunders and C. Donald O'Malley, *Illustrations from the Works of Andreas Vesalius of Brussels* (Cleveland: World Publishing Co.), 28–29, suggest possible identifications for drawings made by Vesalius.
8. Vesalius, *Fabrica,* 273. Similar sketches are *ibid*., 296, 259/359, 307–8/407–8, 323/423. Cf. also his claims for clarity at p. 297/397.

9. *Ibid.*, 275/375. For other drawings of an imagined organ, see pp. 515, 621.
10. *Ibid.*, 289/389; cf. 293/393 for a small drawing of the umbilical veins as found in children.
11. *Ibid.*, 323/423.
12. *Ibid.*, 344/444.
13. *Ibid.*, 258/35, both images, although only the second of them is specifically associated with his lectures. The more elaborate depiction of the tunics of the aorta after dissection, 325/42, may also be his.
14. *Ibid.*, 284/384, said by Saunders and O'Malley 1950, p. 152, to be "probably drawn by Vesalius himself."
15. *Ibid.*, 300; cf. the drawings of the *rete mirabile*, *ibid.*, 621.
16. Monique N. Kornell, "Jan Steven van Calcar (c. 1515–c. 1546), Vesalius's Illustrator," in *Andreas Vesalius and the "Fabrica" in the Age of Printing: Art, Anatomy, and Printing in the Italian Renaissance*, edited by Rinaldo Canalis and Massimo Ciavolella (Turnhout: Brepols, 2018), 99–130, is a detailed study of Calcar's relationship to Vesalius and the *Fabrica,* noted first, albeit imprecisely, by Vasari in the second edition, 1568, of his *Lives of the Painters.*
17. *Ibid*., 121–22.
18. Martin Kemp, "A Drawing for the *Fabrica*: and some Thoughts upon the Vesalius' Muscle-men", *Medical History* 14 (1960), 277–88.
19. Martin R. Caiati et al., "Vesalius' "Philosopher", a Recently Found Drawing by Jan Steven van Calcar," in *In the Shadow of Vesalius,* edited by Rob Van Hee (Antwerp and Apeldoorn: Garant, 2020), 13–24. The figures on successive pages, pp. 163–65, are termed here 'the Plough', 'the Philosopher' and, 'the Afflicted'. The authors report that their scientific studies show that the inserted initials and the drawing were made at the same time.
20. My own suspicions were aroused by the hand, which did not seem to me to be that of Vesalius.
21. François Tortebat, *Abregé d'anatomie, accommodé aux arts de peinture et de sculpture* (Paris: Tortebat, 1668); see also below, p. 173.

22. I am grateful to Monique Kornell for suggesting the link with Tortebat, and to Maurits Biesbrouck for providing me with a larger image of the drawing.
23. Daniel Margócsy et al., *The* Fabrica *of Andreas Vesalius. A Worldwide Descriptive Census, Ownership and Annotations of the 1543 and 1555 Editions* (Leiden and Boston: Brill, 2018), 164; Kornell, "Calcar," 122–23.
24. Kornell, *ibid.,* 122–23, with a bibliography.
25. For the date of the Jena inscriptions, Margócsy et al., *Census,* 93 ("presumably eighteenth-century").
26. G. S. Terence Cavanagh, "A New View of the Vesalian Landscape," *Medical History* 27 (1983), 77–79; John A. Goree, "The Woodblocks of Vesalius and the Printings: From the Renaissance to the Modern Era," *Printing History* 16 (2014), 11–13. The suggestion was first made by Tortebat, see Monique N. Kornell, *Flesh and Bones: The Art of Anatom.* (Los Angeles: Getty Trust Publications) 2022, 8, in his addition of a landscape backdrop to the *Epitome* nudes. Roger Gaskell independently suggested to me that the landscapes might have been drawn by Campagnola. Willy Wiegand, "Marginal Notes by the Printer of the ICONES," in Samuel W. Lambert, Willy Wiegand, and William M. Ivins Jr., *Three Vesalian Essays to Accompany the Icones anatomicae of 1934* (New York: The Macmillan Company, 1952), 39, wondered if the muscle men figures were not intended to form a separate publication; cf. ibid., 39–41.
27. Cf. Michelangelo Muraro and David Rosand, *Tiziano e la silografia veneziana del Cinquecento* (Venice: Neri Pozza, 1976), 123, who believe this portrait can be attributed to him 'con una certa sicurezza'.
28. An early identification of him as a contributing artist to a (now lost) *Last Supper* is very uncertain, Kornell, "Calcar", 110n. 21.
29. Tortebat, *Abregé, Au Lecteur,* 2; Domenico Bonaveri, *Notomie di Titiano* (Bologna, n.p., c. 1688–90); Monique N. Kornell, "A Dating for Domenico Bonaveri's *Notomie di Titiano,*" *Print Quarterly* 33, no. 4 (2016), 379–90.
30. Erika Tietz-Conrat, "Neglected Contemporaary Sources Relating to Michelangelo and Titian," *The Art Bulletin* 25 (1943), 154–59.

31. Patricia Simons and Monique Kornell, "Annibal Caro's After-Dinner Speech (1536) and the Question of Titian as Vesalius's Illustrator," *Renaissance Quarterly* 41, no. 4 (2008), 1069–97.
32. Kemp, "Vesalius"; Kornell, *Flesh and Bones, 2*, judiciously says 'after Campagnola'. A detailed biography of Campagnola is Tobias Benjamin Nickel, *Die Landschaftszeichnungen von Domenico Campagnola (1500–1564)*. Ph.D. Diss., University of Vienna, 2017. For the possibility that he drew the background to the muscle men figures, see above, note 26.
33. Muraro and Rosand, T*iziano,* 129. They discuss the *Fabrica* on pp. 21–30.
34. I am grateful to him for being allowed to present his conclusions here.
35. A succinct account can be found in Mattia Biffis, "Porta, Giuseppe", in *Dizionario Biografico degli Italiani*, vol. 85 (Rome: Treccani, 2001), 98–102; Henry Kaap, "Porta, Giuseppe," in Andreas Beyer, Bénédicte Savoy and Wolf Tegethoff, eds, *De Gruyter Allgemeines Künstler Lexikon, Band 96, Pintaldi – Pretro,* (Berlin: De Gruyter 2017), 368–70.
36. Biffis, "Porta"; idem, *Salviati a Venezia. Un artista immigrato nell'Italia del Cinquecento* (Rome: Artemide, 2021), 63–73.
37. Antonella Imolesi Pozzi, *"L'attribuzione del frontespizio de Le Sorti:* una questione aperta o un falso problema?" in *Un giardino per le arti: "Francesco Marcolino da Forlì," la vita, l'opera, il catalogo. Atti del convegno internazionale di studi, Forlì, 11–13 ottobre 2007*, edited by Paolo Procaccioli, Paolo Temeroli, and Vanni Tesei (Bologna: Compositori, 2009), 269–98.
38. Kaap, "Pozzi," 368–69.
39. Giuseppe Salviati *Regola di far perfettamente col compass la voluta et del capitello ionico et d'ogn' altra sorte* (Venice: Francesco Marcolini, 1552), pref.
40. A suggestion made to me by Monique Kornell.
41. Nickel, *Landschaftszeichnungen,* 87–88.
42. Francisco Guerra, "The Identity of the Artists involved in Vesalius' *Fabrica* 1543," *Medical History* 13 (1969), 37–50.
43. Frank Hieronymus, *Theophrast und Galen—Celsus und Paracelsus. Medizin, Naturphilosophie und Kirchenreform im Basler Buchdruck bis zum dreissigjährigen Krieg* (Basle: Universitätsbibliothek, 2005), 645.

44. His comment to his friend Johannes Gast in 1546 that the larger the book, the greater his pleasure in it is not necessarily to be taken as ironic, O'Malley, *Andreas Vesalius*, 1964: 455, for in his letter to Oporinus, *Fabrica*, viii, he remarked that the *Tabulae* could never be too large for students.
45. Roger Gaskell, "Typography", 50, suggests that Oporinus might have used as many as six presses.
46. Vesalius, *Fabrica,* viii, with the elucidation of Roger Gaskell, "Typography"; the tardy breaking up of the folio page into smaller paragraphs took a surprisingly long time, see Frans A. Janssen, "The Rise of the Typographical Paragraph," in *Cognition and the Book. Typologies and Formal Organisation of Knowledge in the Printed Book of the Early Modern Period*, edited by Klaus A. Enenkel and Wolfgang Neuber (Leiden and Boston: Brill, 2005), 9–32.
47. Matthew Landers, "Early Modern Dissection as a Physical Model of Organization," in *Anatomy and the Organization of Knowledge, 1500–1800,* edited by Matthew Landers and Brian Muñoz (London: Pickering and Chatto, 2012), 11–15, discusses the relations between Vesalius' innovations in anatomy and in printing.
48. Wiegand, "Marginal Notes," 33.
49. This did not prove always possible, perhaps because of differences in the way in which the copy text was set out, see below, pp. 170, 179.
50. Roger Gaskell, "Typography," 43–44. Cf. the later letter to Gast on the desired legibility of the lay-out of the China Letter, O'Malley, *Andreas Vesalius*, 455.
51. Roger Gaskell, "Typography", 37.
52. Vesalius, *Fabrica*, vii.
53. Daniel H. Garrison and Malcolm H. Hast, Andreas Vesalius. *Andreas Vesalius, The Fabric of the Human Body* (Basle: Karger, 2014), LXIX.
54. Above, p. 16–17.
55. Bette Talvacchia, *Taking Positions. On the Erotic in Renaissance Culture* (Princeton University Press, 1999), 166–87. Two plates, *ibid.*, 177–78, 182, showing the womb and female genitalia, were cut entirely anew but incorporating the earlier designs. One image, the 'frog' plate, is mediaeval. See also Kenneth D. Roberts

and John D. W. Tomlinson, *The Fabric of the* Body (Oxford: Clarendon Press, 1992), 170–87.

56. Frank Hieronymus, *Oberrheinische Buchillustration 2, Basler Buchillustration 1500–1545* (Basle: Universitätsbibliothek, 1984), 645. This may also underestimate the extent of Oporinus' part in the decision-making and the expertise of the setters.
57. Vesalius, *Fabrica*, vii.
58. *Ibid.*, viii.
59. Harry Clark, "Foiling the Pirates. The Preparation and Publication of Andreas Vesalius' *De humani corporisi fabrica*," Library Quarterly 51 (1981): 301–11, emphasises the strength of Basle privileges against plagiarists compared with those of Venice.
60. A succinct overview is given by Urs B. Leu, "Die Bedeutung Basels als Druckort im 16. Jahrhundert," in Sven Grosse, Christine Christ-von Wedel and Berndt Hamm, eds, *Basel als Zentrum des geistigen Austauschs in der frühen Reformation* (Tübingen: Mohr Siebeck, 2014), 53–78.
61. Hippocrates, *Libri omnes, ad vetustos codices summo studio collati et restaurati* (Basle: H. Froben and N. Episcopius, 1538).
62. Galen, *Opera omnia, ad fidem complurium & perquam vetustorum exemplariorum ita emendata atque restituta, ut nunc primum nata, atque in lucem aedita, uideri possint* (Basle: Cratander, 1538); Vivian Nutton, *John Caius and The Manuscripts of Galen: Proceedings of the Cambridge Philological Society,* Supplementary Volume 13 (1987), 43–44; Hieronymus, *Theophrast und Galen,* 87–89; Beate Gundert, "Zu den Quellen der Basler Galen-Ausgabe (1538)," in *Ärzte und ihre Interpreten: Medizinische Fachtexte der Antike als Forschungsgegenstand der Klassischen Philologie,* edited by Carl W. Müller, Carl Brockmann and Carl W. Brunschön (Munich and Lepizig: K. G. Saur, 2006), 81–100.
63. Goree, "Woodblocks", 7–9; above, pp. 41–42. See also below, pp. 118–19, for possible borrowings of ideas from this work, whether adopted or rejected.
64. Gemusaeus 1543, preface; German translation in Hieronymus, *Theophrast und Galen*, 198; Christoph Reske, *Die Buchdrucker des 16. und 17. Jahrhunderts im deutschen Sprachgebiet. Auf der Grundlage des gleichnamigen Werke von Josef Benzing, 2. überarbeitete und erweiterte Auflage* (Wiesbaden: Harrassowitz, 2015),

78–80; Martin Steinmann, *Johannes Oporinus, ein Basler Buchdrucker um die Mitte des 16. Jahrhunderts* (Basle and Stuttgart: Helbing & Lichtenhahn, 1967), 29–32.

65. Reske, *Buchdrucker,* 78–80, citing a letter of Oporinus to Joachim Camerarius confessing that he was so afraid of the continued costs of printing with Winter that he hoped to have his own press.
66. Steinmann, *Oporinus,* 32.
67. Steinmann, *ibid.*: 41–2.
68. *Ibid.*, 37: Vesalius, *Fabrica,* viii.
69. Vesalius, *Fabrica,* viii. For the possibility that he was also helped by the wealthy Augsburg patrician Wolfgang Herwarth in contributing financially to the printing costs, *ibid.*, 35.
70. Vesalius, *Fabrica,* vii.
71. O'Malley, *Andreas Vesalius,* 455, n. 149.
72. Sachiko Kusukawa, *Picturing the Book of Nature. Image, Text and Argument in Sixteenth-Century Human Anatomy and Medical Biology* (Chicago: Chicago University Press, 2012), 32–35. Jacqueline Vons and Velut, Stéphane, *André Vésale. Résumé de ses livres sur la Fabrique du corps* humain (Paris: Les Belles Lettres, 2008), LXXVIII, describes the paper used for the *Epitome* as "thick"; watermarks show it was produced in Basle. Paper costs may have amounted to more than 70% of Oporinus' expenses.
73. Steinmann, *Oporinus,* 37, cites the preface to a later printing of 1543 as the first indication that Oporinus was seeking financial help at least in the short term. Winter's continuing friendship with Vesalius may indicate that he was happy not to participate in this venture, and that Vesalius concurred with his decision.
74. Kusukawa, *The Book of Nature,* 200 and 285, n. 5, noting that the cost of the copy owned by Gast's friend is recorded as 10 *pfund*, giving, *ibid.*, n. 7, a florin as the equivalent of 2 pfund.
75. Albrecht Burckhardt, *Geschichte der medizinischen Fakultät zu Basel 1460–1900* (Basle: Friedrich Reinhard, 1917), 15, cf. 39 for Thorer paid 40 pf. These sums would be increased by those for private teaching and practice.
76. Margócsy et al., *Census,* 9. For suggestions for the print runs of other Vesalian works, see Stephen N. Joffe and Veronica Buchanan, "The Andreas Vesalius Woodblocks, from Creation to Destruction," *Acta Medicinae Historiae Adriaticae* 14, no. 2 (2016), 371.

77. Margócsy et al. 2018: 13.
78. *Ibid.,* 7–8, 13, 31–35,
79. Vesalius, *Fabrica,* vi–viii. His plea was successful, and the royal privilege was recorded on the title page.
80. Sachiko Kusukawa, *Andreas Vesalius. Anatomy and the World of Books* (London: Reaktion Books, 2024), 9. The dedication copy was sold to a private bidder at Christie's, London, on March 18, 1998: Margøcsy *et al., Census,* I/A8; p. 457.
81. His list of names contains many who had studied at Alexandria, but by Galen's day human anatomy there had long been confined to observing a skeleton and possibly a foetus, and dissection and vivisection were possible only on animals, Claire Bubb, *Dissection in Classical Antiquity: A Social and Medical* History (Cambridge, Cambridge University Press, 2022, 95–120); Vivian Nutton, *Ancient Medicine,* ed. 3, (London and New York, Routledge), 2024, 186.
82. Vesalius, *Fabrica,* iii, 81. For the role of later summarisers, 572.
83. *Ibid.,* 300; cf. 245/345 for Galen's erroneous criticism of earlier authorities. This mistaken belief in a long pre-Galenic tradition he repeated in *Epistola Chynae,* 95; trans. Garrison, *China Root,* 114.
84. Below, pp. 159–60, 163.
85. Cynthia Klestinec, "Juan Valverde de (H)Amusco and Print Culture. The Editorial Apparatus in Vernacular Anatomy Texts," in *Zergliederung – Anatomie und Wahrnehmung in der Frühen Neuzeit,* edited by Albert Schirrmeister and Matthias Pozsgai (Frankfurt am Main: Vittorio Klostermann, 2005), 83–87.
86. Or possibly a copy of the book itself, which is undated.
87. Andreas Vesalius, *Anatomicarum Gabrielis Falloppii observationum examen* (Venice: Francesco de' Franceschi, 1564), 71. For Canani's book, Giovanni Gliozzi,"Canano, Giovanni Battista, il Giovane," in *Dizionario biografico degli italiani* (Rome: Treccani, vol. 17 (1974)): 714–16; Marco Bresadola, "*Giovanni Battista Canani: l'umanesimo medico e la riforma dell'anatomia rinascimentale*", *Schifanoia* 50/51 (2016), 13–20; Francis Van Glabbeek et al., *Myology before Vesalius. Giovanni Battista Canani (1515–1579 n.s.) and his* Musculorum Humani Corporis Picturata Dissectio (Antwerp and Apeldoorn: Garant, 2023). His

claim in December 1543 for the payment of money wrongly withheld from him in the previous year during his absence is imprecise: some teaching in the first term at Padua cannot be excluded.

88. O'Malley, *Andreas Vesalius,* 136.
89. Vesalius, *Fabrica,* 19–20; Maurits Biesbrouck, and Omer Steeno, "Andreas Vesalius (1514–1564) and Christophorus Pfluegel (1520–after 1587) and the Role of the Students in the Siege of Leuven in 1542," in *In the Shadow of Vesalius,* edited by Rob. Van Hee (Antwerp and Apeldoorn: Garant, 2020), 24–35.
90. By mistake a setter numbered pp. 313–491 as 213–391. I give both numbers to avoid confusion.
91. Vesalius, *Fabrica,* 335/435. For the insertion at p. 329/429, see below, pp. 120–21, 124–25.
92. Setting new illustrations as separate images in the text was, clearly, difficult at this stage. For the significance of these initials, see below, pp. 120–21, 124–25.
93. O'Malley, *Andreas Vesalius,* 137–38.
94. Andreas Vesalius, *Suorum de humani corporis fabrica librorum epitome* (Basle: J. Oporinus, 1543); Levi R. Lind, *The* Epitome *of Andreas Vesalius* (New York: Macmillan Co, 1949).; Vons and Vélut, *Résumé.*
95. Vesalius, *Fabrica,* v recto.
96. Andrea Carlino, *Paper Bodies. A Catalogue of Anatomical Fugitive Sheets 1538–1687, Medical History Supplement* 19 (London: The Wellcome Institute, 1999).
97. Vons and Vélut, *Résumé,* LXXVII–LXXX; Goree, "Woodblocks", 16–17, points to a slight deterioration in one of the woodcut blocks between the printing of the *Epitome* and that of the *Fabrica.*
98. Vons and Vélut, *Résumé*; for the Cambridge coloured copy of the *Epitome*, Zoltan I. Pozeg and Eugene S. Flamm, "Vesalius and the 1543 Epitome of his "*De humani corporis Fabrica librorum*": A Uniquely Illuminated Copy," *Papers of the Bibliographical Society of America* 103 (2009), 199–220.
99. Carlino, *Paper Bodies,* 118–98.
100. Frank Hieronymus, "Physicians and Publishers in Sixteenth-century Basle," in *The German Book 1450–750. Studies Presented to David L. Paisey on his Retirement,* edited by John L. Flood and William A. Kelly (London: The British Library, 1995), 95–97.

101. Alban Thorer, *Abupetri Rhazae Maomethi ....summi medici opera exquisitiora.Per Gerardum Toletanum medicum Cremonensem, Andream Vesalium Bruxellensem, Albanum Torinum Vitoduranum Latinitate donata* (Basle: Heinrich Petri, 1544); one may speculate on the relationship of this edition to the paraphrase of the whole work that Vesalius claims to have destroyed on entering imperial service, below, pp. 157–58.
102. Vesalius, *Von der menschen Cörpers Anatomey* (Basle, J. Oporinus, 1543).
103. Henry E. Sigerist, "Albanus Torinus and the German Edition of the "*Epitome*" of Vesalius," *Bulletin of the History of Medicine* 14 (5), 1943, 663–65, translates the last part of the preface.
104. Burckhardt, *Geschichte,* 39–42.
105. Sigerist, "Albanus Torinus."; Hieronymus, Buchillustration, 646–67; Vons and Vélut, *Résumé*, LXXX. Less familiar is the Yiddish translation of the *Epitome*, copied around 1590–5 in Hesse, based on Thorer's version, University of Pennsylvania, Kizlak Center for Special Collections, Manuscripts LJS 485.
106. Jonathan Sawday, *The Body Emblazoned. Dissection and the Body in Renaissance Culture* (London: Routledge, 1995), 66–76; Carlino, *Books of the* Body, 6–68; Kusukawa, *Andreas Vesalius.*
107. Vivian Nutton, *Renaissance Medicine,* 110, Fig. 4, 2; Mary-Jo Nelson, *An Analysis of Sixteenth-century Venetian Title Pages of Medical Books* (Ph.D. Diss., University of Kansas, Laurence, KA., 1976), 351–58, 594–97.
108. Jerome J. Bylebyl, "Interpreting the *Fasciculo* Anatomy Scene," *Journal of the History of Medicine and Allied Sciences* 45 (1990), 285–316. Carlino, *Books of the Body*, 8–68, and McCall, *The Art of Anatomy* (London: Reaktion Books, 2023), 57–76, discuss earlier dissection illustrations.
109. Nutton, *Renaissance Medicine*, 110, Fig. 4.2. The later Juntine editions show Galen, as described in *On Prognosis,* dissecting an animal before a select audience, some of whom are named.
110. Ruben Eriksson, *Andreas Vesalius' First Public Anatomy at Bologna 1540. An Eye-witness Report by Baldasar Heseler medicinae scolaris together with his Notes on Matthaeus Curtius' Lectures on Anatomia Mundini.* (Uppsala and Stockholm: Almqvist and Wiksell, 1959), 85, although the round number may simply indicate a very large crowd.

111. Saunders and O'Malley, 248–51. The alternative, that these were versions copied from the published frontispiece, is less convincing.
112. David Soulier, "The Anatomy Theatre in Renaissance Italy: a Space of Justice?" in *The Spaces of Renaissance Anatomy Theater*, edited by Leslie R. Malland. (Wilmington: Vernon Press, 2022), 27, stressing that the figure stands in a position of honour at the right of the dissector.
113. Renatus Henerus, *Adversus Jacobi Sylvii depulsionum anatomicarum calumnias, pro Andrea Vesalio apologia* (Venice: n.p., 1555), 116v–117v, gives a selection of the names of many of the audience at his dissection at Bologna in late January 1544. For some wealthy non-medics attending his lectures, Ludovico Panizza, *De venaesectione in inflammationibus* (Venice: F. Camozzi, 1561), 97r. The Bologna statutes of 1405 specified the maximum numbers of attendees at 30, David Lines, *The Dynamics of Learning in Early Modern Italy* (Cambridge, MA, and London: Harvard University Press, 2023), 89, but numbers at the end of the next century were much greater, Giovanna Ferrari, "Public Anatomy Lessons and the Carnival; the Anatomy Theatre of Bologna," *Past and Present* 117 (1987), 50–106; *eadem,* "Tra medicina e chirurgia: la rinascita dell'anatomia e la dissezione come *spettacolo,*" in *Il Rinascimento italiano e l'Europa: Volume quinto, Le Scienze.* edited by Antonio Clericuzio and Germana Ernst (Rome: Fondazione Cassamarca, 2008), 41–68.
114. The accuracy of some of the details, e.g. the table and the scalpel, suggests that the artist had himself attended such a demonstration by Vesalius.
115. How this structure should be constructed is described by Alessandro Benedetti, *Historia corporis sive Anatomice* (Venice: B. Guerraldo, 1502), sigg. a iiv-iiir, who specifies that it should be erected a large airy place in winter.
116. Vesalius, *Fabrica,* 548, although, as Kusukawa, *Andreas Vesalius*, 12, points out, the architectural setting does not correspond to any building in the two cities, and Vesalius may simply be referring to the temporary wooden seating. Sawday, *The Body Emblazoned*, 66–74, offers some suggestive ideas about the role of this 'tempietto'.
117. Paula Findlen, "Anatomy Theatres, Botanic Gardens, and Natural History Collections," in *The Cambridge History of Science, vol.*

3, *Early Modern Science*, edited by Katharine Park and Lorraine Daston (Cambridge: Cambridge University Press, 2006), 272–89; Kusukawa, *Andreas Vesalius,* ch. 6, develops the notion of the theatre as a place for seeing.

118. Andrew Cunningham and Tamara Hug, *Focus on the Frontispiece of the* Fabrica *of* Vesalius (Cambridge, Wellcome Unit for the History of Medicine, 1994), 37–39; Kusukawa, *Andreas Vesalius*, 12–19.
119. He is, of course, mentioned in the colophon. Wiegand, "Marginal Notes," 1952: 35, doubts that this refers to Oporinus, suggesting it might be the initials of the artist, but this is unlikely. This use of initials would be a learned *concetto,* accessible only to those of sufficient intelligence.
120. Katharine Park, *Secrets of Women: Gender, Generation and the Origins of Human Dissection* (London: Zone Books, 2006).
121. Benedetti, *Anatomice,* sig a ii v, recommending that one should choose corpses of unknown and humble origin from far away to avoid causing offence to neighbours and relatives.
122. London, Wellcome Collections, WMS 5265.
123. Helen King, *The One-sex Body on Trial: The Classical and Early Modern Evidence,* (Farnham and Burlington, VT: Ashgate, 2013), 51–62.
124. Vesalius, *Fabrica,* 533, 539, presumably dissected in 1541/2.
125. Vivian Nutton, "Wittenberg Anatomy," in *Medicine and the Reformation,* edited by Ole P. Grell and Andrew Cunningham (London and New York: Routledge, 1993), 111–32.
126. Sawday, *The Body Emblazoned,* 70–71, also noting the parallel between the surroundings of the dissection and the womb and the 'tempietto' around the foetus.
127. Bylebyl, "The *Fasciculo Scene*", 51–53; Carlino, *Books of the Body,* 49–52; Cynthia Klestinec, *Theaters of Anatomy: Students, Teachers and Traditions of Dissection in Renaissance Italy* (Baltimore: Johns Hopkins University Press, 2011), 32–34, who suggests that the figure of the skeleton takes the place of the *lector,* while the *dissector* grumbles under the table.
128. The putti appear in the earliest version of the frontispiece, which may suggest that Vesalius had by then already decided upon the overall design for the initials.

129. I cannot make out the significance of the two gargoyle-like heads at the top of the cartouche.
130. Easiest accessible in Saunders and O'Malley, *Illustrations,* 42–43.
131. Above, p. 99.
132. As something made, it may also allude to the work of the divine creator, often stressed throughout the volume.
133. Jackie Pigeaud, "Formes et normes dans le *De Fabrica de Vésale,*" in *Le corps à la Renaissance. Actes du XXX*[e] *Colloque de Tours,* edited by Jean Céard, Marie-Madeleine Fontaine and Jean-Claude Margolin (Paris: Aux Amateurs de Livre, 1990), 400–402; Andrea Carlino, "Medical Humanism: Rhetoric and Anatomy at Padua, circa 1540," in *Rhetoric and Medicine in Early Modern Europe,* edited by Stephen Pender and Nancy S. Struever (Farnham and Burlington, VT: Ashgate, 2012), 111–28.
134. Theophilus Protospatharius, *De corporis humani fabrica libri quinque a Junio Paulo Crasso Patavino in Latinam orationem conversi* (Venice: O. Scotus, 1536), although Crassus may well have borrowed from Cicero. For Theophilus, see Petros Bouras-Vallianatos, "Galen in Byzantine Medical Literature," in *Brill's Companion to the Reception of Galen,* edited by Petros Bouras-Vallianatos and Barbara Zipser (Leiden and New York: Brill, 2019), 88–89.
135. One of the rare instances of the use of '*fabrica*' in the text, Vesalius, *Fabrica,* 506–507, comes immediately after the sole mention of Theophilus, described as "certainly not an anatomist".
136. For author portraits in general, Nelson, *Title Pages,* 351–58; Ruth Mortimer, "The Author's Image: Italian Sixteenth-century Printed Portraits." *Harvard Library Bulletin* 7 (2) (Summer 1996), 7–87. The old discussion of the progeny of this portrait by M. H. Spielmann, *The Iconography of Andreas Vesalius, Research Studies in Medical History* 3 (London: John Bale, Sons, & Danielsson, Ltd, 1925), is still valuable.
137. Alessandro Achillini, *Anatomicae annotationes magni Alexandri Achillini Bonon* (Bologna: H. De Benedictis, 1520), t.p.; Leonhart Fuchs, *De historia stirpium* (Basle: M. Isingrin:1542), sig. α iv; Kusukawa, *The Book of Nature,* 107–109, with Fig. 5.1.
138. Nutton, *Renaissance Medicine,* 168. A good selection of images of physicians and surgeons is in Kitti Jurina, *Vom Quacksalber zum Doctor Medicinae* (Cologne: Bohlau Verlag, 1985), 186–205.

139. Kornell, *Flesh and Bones,* 126, no. 16, Fig. E; below, notes 407, 503.
140. The title page of his printed *De re anatomica* (Venice: N. Bevilacqua, 1559) shows him dissecting a corpse and wearing a more day-to-day outfit.
141. His age, 28, and the year, 1542, appear on the front of the table. The slightly over-large head is shown also in the title page, and it is hard to believe that so scrupulous an observer as Vesalius would have allowed this image if he had thought it distorted or exaggerated.
142. For the unusual choice of a female arm, Carlino, *Books of the Body,* 52.
143. In Colombo's portrait his white hands, and particularly the right, also attract the viewer's attention, see the images on the Tatton Park and National Trust websites, Tatton Park, NT 1298173.
144. A comment made to me by a former hand surgeon, Rev. Lore Chumbley.
145. William Schupbach, *The Paradox of Rembrandt's 'Anatomy of Dr. Tulp', Medical History, Supplement* 12 (London: The Wellcome Institute for the History of Medicine, 1982), 18 and Appendix 1, gives a selection of texts. In all instances, man or mankind is resolutely masculine, which adds to the uniqueness of Vesalius' image.
146. O'Malley, *Andreas Vesalius,* 148–49, discusses the wording on the scroll, suggesting that its reference to the arm might refer to a comment in Vesalius' destroyed *Notes on Galen,* but other explanations are possible.
147. For Vesalius' repetition of this passage, with a possible hit at Galen's self-centredness, *Fabrica,* 329/429.
148. Galen, *On Mixtures* I,9: I,563–64K., translated by Peter N. Singer and Philip J. Van der Eijk, *Galen, Works on Human Nature,* vol. 1, *Mixtures (De* temperamentis) (Cambridge: Cambridge University Press, 2018), with valuable notes.
149. Niccolò Massa, *Anatomiae liber introductorius* (Venice: F. Bindoni and M. Pasini, 1536), f. 97r.
150. Nancy G. Siraisi, *Medicine and the Italian Universities, 1250–1600* (Leiden: Brill, 2001), 157–62; Pablo Maurette, "The Organ of Organs: Vesalius and the Wonders of the Human Hand,"

The Journal of Medieval and Early Modern Studies 48 (2018), 79–104.

151. It is a moot point whether the size of the dissected figure is an allusion to the grandeur of ancient bodies, as Sylvius later argued in his attack on Vesalius.
152. Vesalius, *Fabrica,* 14, 16.
153. John Caius, *Libri aliquot graeci* (Basle: H. Froben and N. Episcopius, 1544), 299; *idem, De libris propriis* (London, W. Seres, 1570), 11r: tr. Nutton, *Autobibliography*, 60–2; Vesalius, *Fabrica* 1555: 17.

    Vesalius, *Fabrica,* 1543: 60, 71, 73.
154. Vesalius, *Ibid.,* 17.
155. Vesalius, *Ibid.*, 36. Cf. 621, an image deliberately introduced to show the arteries by the pituitary gland as well as Vesalius' ability to tell human from animal anatomy.
156. This section depends on Jacqueline Vons, 'Les squelettes de Vésale,' In Vons, J. (ed.), *La* Fabrique *de Vesale. La mémoire d'un livre* (Paris: Bibliothèque interuniversitaire de Santé, 2008), 37–64.
157. Since Virgil died in 19 BCE, he cannot have written a poem on the death of Maecenas, who died eleven years later.
158. O'Malley, *Andreas Vesalius*, 149, devotes a mere twelve lines to them and misrepresents Vesalius' novelty. Described at length in Samuel W. Lambert, "The Initial Letters of the Anatomical Treatise *De Humani Corporis Fabrica* of Vesalius," in Lambert et al., *Three Vesalian Essays*, 1–24. The most recent study of them is Guy Gilias, "Ornamental letters in the *Fabrica* 1543 and 1555," in *Art of Vesalius,* edited by Rob Van Hee (Antwerp and Apeldoorn: Garant, 2014) 97–112, descriptive but with good clear illustrations.
159. Sarah R. Kyle, *Medicine and Humanism in late Medieval Italy: the Carrara Herbal in Padua* (London: Routledge, 2017).
160. Vivian Nutton, "Medicine: the Dresden Galen," in *Bild und Text im Mittelalter,* edited by Karin Krause and Barbara Schellewald (Cologne, Weimar and Vienna, Böhlau Verlag, 2011), 347–361.
161. Francesca Petrucci Nardelli, *La lettera e l'immagine: le iniziali "parlanti" nella tipografia italiana* (*secc. XVI–XVIII* (Florence: Olschki, 1991), 109–201.

162. Steinmann, *Oporinus*, 36, 41. For historiated frontispieces, see above, p. 116.
163. Gustav Schneeli and Paul Heitz, *Initialen von Hans Holbein* (Strasbourg: J. E. Ed. Heitz, 1900), 12–14.
164. Hieronymus, *Theophrast und Galen,* 214–15, 196; Galen, *Opera omnia,* 1538, above, p. 41, with Hieronymus, *Theophrast und Galen,* 192; Galen, *Opus medicum practicum* (Basle: H. Froben and N. Episcopius, 1537).
165. Giovanni Manardo, *Epistolae medicinale* (Basle: J. Bebel, 1535); Hieronymus, *Theophras und Galent,* 391, 393; I rely for examples on the images in this book.
166. Guinter, Institutiones 1536: Hieronymus, *Theophrast und Galen,* 285.
167. Hieronymus, *Theophrast und Galen*, 184; Alexander of Aphrodisias, *De causis febrium* (Basle: Robert Winter, 1542) with Hieronymus, E.79–81. Paul of Aegina, *Libri Septem,* (Basle: Andreas Cratander, 1538) with Hieronymus, *Theophrast und Galen,* 196.
168. Koran, M*achumetis saracenorum principis eiusque successorum vitae, ac doctrina, ipseque alcoran, … quae … Petrus abbas Cluniacensis … ex Arabica lingua in Latinam transferri curavit,* (Basle: J. Oporinus), respectively p. 8; sigg. α 2r, α 3v. The latter is a version of that used by Cratander in 1538, Hieronymus, *Theophrast und Galen,* 196.
169. Bassiano Landi, *De humana historia, uel singularum hominis partium cognition.* (Basle: J. Oporinus, 1542), respectively p. 1, sig. α 2r, p. 98.
170. Giuseppe Ongaro, "Il *'De humana historia'* (1542) di Bassiano Landi," *Atti della VI Biennale della Marca per la storia della medicina (Fermo, 29 aprile-2 maggio, 1965,* (Fermo: n.p., 1965), 265–78; Fabrizio Bigotti, "Commenting on Aristotle with a Knife. The Heretical Anatomies of Bassiano Landi," *NTM, n.s.* 31 (2023), 1–26.
171. Landi, *De humana historia*, 3, 112. The allusion to Vesalius was first noticed by Giuseppe Ongaro, "Bassiano Landi e Andrea Vesalio," *Atti e memorie dell'Accademia patavina di scienze, lettere ed arti* 110, part 2 (1997–8), 43.
172. E.g. Schneeli and Heitz, *Initialen*, 12–14 and pl. XXV, LXIX–LXX.

173. John A. Goree, “Woodblocks”, 7–9; Fuchs, *De historia stirpium*; Frederick G. Meyer, Emily E. Trueblood, and John L Heller (eds), *The Great Herbal of Leonhart Fuchs: De historia stirpium commentarii insignes, 1542* (Stanford: Stanford University Press and Cambridge: Cambridge University Press, 1999); Hieronymus, *Theophrast und Galen*, 238–56.
174. Francesco Marcolini, *Le sorti* (Venice: Francesco Marcolini, 1541).
175. This comes from a 1523–4 series, Schneeli and Heitz, *Initialen*, Number XXXV, pl. LXX. The identification was first made by Robert Herrlinger, “Die Initialen in der <Fabrica> des Andreas Vesalius,” *Acta Medicae Historiae Patavina* 10, no. 1 (1963–4), 114. For the use of another letter (A) from this series in 1548–9, see below, p. 165.
176. Roger Gaskell suggests that it might have been inserted by a workman, perhaps noting the relevance of a familiar initial to the chapter, but, even so, its repetition in 1555 indicates that it was not considered inappropriate.
177. Four, C, F, H and L, show humans and in another, O, a hooded human figure hands over a recently executed corpse to *putti.*
178. The fact that what is going on several of these images is unclear to the non-medic or ambiguous matters less than the overall message that such activity is fun and accessible to a learned reader.
179. Vesalius, *Fabrica,* I r.
180. Talvacchia, *Taking Positions.*
181. Vesalius, *Fabrica,* 155: letter C, first used on p. 3; p. 162 letter P, first used on p. 62; see also Ongaro, ‘Bassiano Landi’, 44–51, for a discussion of letter C and the choice of images.
182. Above, p. 118.
183. Adolf Faller, As iniciais da II ª ediçião (1555) de obra ‘De humani corporis fabrica Lib;ri septem’ da Andrea Vesalio (Rio de Janeiro: CIBA, 1951), 13–14.
184. Wiegand, “Marginal Notes”, 38. Detailed description of the initials in Lambert, “Initial Letters”. At what stage decisions about font and spacing were taken is unclear, but artist, cutter and typesetter must all have been given appropriate instructions by Oporinus after consultation with Vesalius.

185. Bassiano Landi, *Iatrologia* (Basle: J. Oporinus, 1543), sig. a2r and pp, 1 and 79; Vesalius, *Epistola Chynae*, 1546, sigg. A 2r, B 2r; pp. 3, 200, 201 tr. Garrison.
186. The English version by Richardson and Carman appeared in five volumes between 1998 and 2009, that by Garrison and Hast in 2014, incorporating also a version of the changes in 1555 and the most important ones in the later annotations.
187. Although Daniel Garrison, *The Fabric of the Human Body* (Karger: Basle, 2014), LXVIII wonders whether a corrector at the press was employed to 'improve' the Latin. For his 1552 publisher's view of his Latin, see below, p. 231, note 186.
188. Vesalius, *Fabrica*, 208.
189. William F. Richardson, *Vesalius, On the Fabric of the Human Body,* translated by William F. Richardson and John B. Carman (Novato, CA: Norman Publishing, 1998–2009), vol. I, xxiv–xxx.
190. Landi, *De humana natura,* 3.
191. For his decision to avoid earlier technical terms, Vesalius, *Fabrica*, 272/372.
192. Richardson, *Vesalius,* III, xii–xiii.
193. Lind, *The* Epitome; *idem, Studies in pre-Vesalian Anatomy.*
194. Eriksson, *Heseler*; Nutton, *Johann Guinter*; Klestinec, *Theaters*; Michael Stolberg, *Gabrielle Falloppia 1522/3–1562. The Life and Work of a Renaissance* Anatomist (London and New York: Routledge, 2023); two essays by Michael Stolberg. "Learning Anatomy in Late-sixteenth Century Padua," *History of Science* 56 (2018), 381–402, and "Teaching Anatomy in Post-Vesalian Padua: An Analysis of Student Notes," *The Journal of Medieval and Early Modern Studies* 48 (1) (2018), 61–77, provide useful information from student notes on others' lectures from the 1550s and later.
195. Bresadola, "Canani"; Van Glabbeek et al., *Canani.*
196. Landi, *De humana natura,* 85: the Preface explains the role of the Creator; see also Silvia Ferretto, *Maestri per il metodo di trattar le cose. Bassiano Lando, Giovan Battista da Monte e la scienza della medicina nel xvi* secolo (Padua, CLEUP, 2012); and the references, above, note 408.
197. Stolberg, *Gabrielle Falloppia*; "Learning Anatomy"; "Teaching Anatomy",

198. Luis A. Salas, *Cutting Words: Polemical Dimensions of Galen's Anatomical Experiments*. Leiden and Boston: Brill, 2020; D*issection*, 83–91.
199. Vesalius, *Fabrica,* Pf. Ii–iv, mentions Paris, Padua and Bologna, noting the enthusiasm of younger Parisians.
200. Vesalius, *Fabrica,* 233; at *Fabrica,* 548 he refers to Giorgio Valla's work on medical terminology, above, p. 31, but does not regard him as an anatomist.
201. Vesalius, *Fabrica,* 42, 495. For the name of the Cardinal, wrongly identified by O'Malley, *Andreas Vealius,* 1964: 130, followed by Richardson and Garrison, see above, p. 128.
202. Vesalius, *Fabrica,* 56. David Soulier by letter suggests a discreet allusion to Benedetti's dedicatory epistle in the Preface.
203. Sylvius is described in the Preface, p. iii, as "someone who can never be praised too highly", but we are never told what Vesalius learned from him.
204. Vesalius, *Fabrica,* 83, 514.
205. *Ibid.,* 81, 282/382; 600, 640–41 for their incompetent teleology.
206. *Ibid.,* 600, 278/378, 276/376,
207. *Ibid.,* 94.
208. *Ibid.,* 322/422, lumping together supporters of both sides of the argument.
209. *Ibid., Fabrica,* 637.
210. Juan José Barcia Goyanes, *El mito de Vesalio* (Valencia: Universidad de Valencia, 1994).
211. Vesalius, *Epistola Chynae*, 196–7: tr. Garrison, *China Root,* 230; for his grandfather's manuscricpt of the *Ad Almansorem*, above, pp. 6–7.
212. Véronique Boudon-Millot, "Vésale lecteur de Galien et d'Hippocrate," in *La* Fabrique *de Vésale. La mémoire d'un livre,* edited by Jacqueline Vons (Paris: Bibliothèque interuniversitaire de Santé, 2016), 13–16.
213. Vesalius, *Fabrica,* 591.
214. *Ibid.,* 144, but in doing so, they would also see the errors of Aristotle and Galen, 144–45; cf. also 161; Andrew Cunningham, "Fabricius and the 'Aristotle Project' in Anatomical Teaching and Research in Padua," in *The Medical Renaissance of the Sixteenth Century,* edited by Andrew Wear, Roger French, and

Iain M. Lonie (Cambridge: Cambridge University Press, 1985), 195–222.

215. Vesalius, *Fabrica,* 55, 309.
216. *Ibid.,* 219/319.
217. *Ibid.,* 85, 265, 243–4/343–4.
218. *Ibid.,* 83; 114, 218/318, 526, 536, 540.
219. *Ibid.* 582, 237–8/327–8.
220. *Ibid.* 256/356. Sarah Parker, "Subtle Bodies; The Limits of Categories in Girolamo Cardano's *De Subtilitate,*" in *Anatomy and the Organization of Knowledge, 1500–1800,* edited by Matthew Landers and Brian Muñoz (London: Pickering and Chatto, 2012), 71–84, contrasts Vesalius' and Estienne's traditional Galenist teleology with Cardano's much looser application.
221. Vesalius, *Fabrica,* 636, 641.
222. *Ibid.,* 65.
223. *Ibid.,* I, praising his work in therapeutics and surgery, but not anatomy.
224. *Ibid.,* 9,
225. And indeed Galen, see Salas, *Cutting Words,* and Bubb, *Dissection,* 83–91, for Galenic strategies.
226. Vesalius, *Fabrica,* 100. For his doubts about the bone in the heart, *ibid.,* 93–4.
227. *Ibid.,* 310.
228. *Ibid.,* 265–66, 278/378, 282/382.
229. *Ibid.,* 265, 286, 299, 533.
230. *Ibid.,* 278/378.
231. *Ibid.,* i, iv, 83, 245/345, 640.
232. O'Malley, *Andreas Vesalius,* 149, cf. 145 for his criticism of Paris anatomy.
233. For Vesalius' attitude to anomalies, see Siraisi, *Medicine and the Italian Universities,* 287–327.
234. Vesalius, *Fabrica,* 44, 92—although the Roman writer Celsus, *On Medicine* VIII, 1,15, also describes three bones.
235. *Ibid.,* 67. The whole section on the names of bones, *ibid.,* 7–10, confirms Vesalius in his view that there were teachers of human anatomy in pre-Alexandrian Greece, and cf. p. i for his use of a similar claim by Galen.

236. Nutton, *John Caius and the Manuscripts of Galen,* 96. Modern scholars side more with Vesalius in their appreciation of Galen's anatomical subjects.
237. Vesalius, *Fabrica,* 44; *Epistola Chynae,* 194.
238. Vesalius, *Fabrica,* 661; Eriksson, *Heseler.*
239. Carlino, *Books of the Body,* above, p. 52.
240. Cf. Stolberg, *Falloppia,* 37–42.
241. Niccolò Massa, *Epistolae medicinales et philosophicae* (Venice: F. Bindoni and M. Pasini, 1550), 52v, 58v. Massa himself blamed his other commitments. See also Roth, *Andreas Vesalius,* 1892), 183–87.
242. John Burd Carman's summary accounts of the anatomical significance of each book that precede each volume of the Richardson translation provide an excellent introduction for the non-anatomist.
243. Andreas Vesalius, *De humani corporis fabrica* (Basle: J. Oporinus), 42–44; 87 and 91; 94.
244. *Ibid.,* 125–26; 142; 142.
245. Ibid., 246; 274–5, 282–3, 293–94.
246. *Ibid.,* 276/376. 590; cf. O'Malley, *Andreas Vesalius* 166–68.
247. Vesalius, *Fabrica,* 324–5/424–5; 512. One of his autopsies that showed the diseased spleen may have been attended by his teacher Da Monte, Giovanni Battista Da Monte, *In primi libri Canonis Avicennae primam Fen, profundissima commentaria* (Venice, V. Valgrisi and B. Constantino, 1558), 642 = *Medicina universa* (Frankfurt, V. Marnius and J. Aubrius, heirs of A. Wechel, 1587), 508.
248. Vesalius, *Fabrica,* 642, 622; Douglas J. Lanska, "Evolution of the Myth of the Human Rete Mirabile Traced through Text and Illustrations in Printed Books: The Case of Vesalius and his Plagiarists," *Journal of the History of the Neurosciences,* 31(2022), 221–61. Cf, also for his work on the brain, Charles Singer, *Vesalius on the Human* Brain (London: Oxford University Press for the Wellcome Historical Medical Museum, 1952; Stéphane Velut, "L'encéphale et ses enveloppes," in *La* Fabrique *de Vesale. La mémoire d'un livre,* edited by Jacqueline Vons (Paris, Bibliothèque interuniversitaire de Santé, 2016), 84–104; Marco Catani and Stefano Sandrone, *Brain Renaissance from Vesalius to Modern* Neuroscience (Oxford: Oxford University Press, 2015), 43–182.

CHAPTER 4

After 1543

The Parting of the Ways and First Reactions

Vesalius did not stay long in Basle once the *Fabrica* had been printed. Soon after he had stood as godfather to Hans Rudolf Winter, the son of his former publisher, on August 2, 1543, he left to travel to Speyer to deliver in person the presentation copies of the *Fabrica* and the *Epitome* to the Emperor Charles V. He was not to know that Charles had moved North to Mainz to be closer to the scene of impending fighting against the Duke of Cleves and his French backers further down the Rhine. It is not clear whether it was at this point or a little later that he brought with him Oporinus' stepson Jacob in the hope of finding him a suitable post in the echelons of the court. If so, it was a failure, for after an initial period there, Jacob reverted to his old ways, eventually being executed for his crimes in 1547.[1] Vesalius' interview with the Emperor was more successful. He was graciously thanked and was quickly enrolled as one of the physicians in the imperial household.

At this stage he had not committed himself fully to imperial service, if we can believe what he says in his *China Root Letter*.[2] His library and much else was still in Padua, and he had many projects envisaged, if not already begun. They included a shortened version of the *Fabrica*, a complete paraphrase of Rhazes' *Ad Almansorem*, a work on practical medicine, another on surgery and the publication of a huge collection of

V. Nutton, *Andreas Vesalius and his Fabrica, 1537–1564*, Palgrave Studies in Medieval and Early Modern Medicine,
https://doi.org/10.1007/978-3-031-69565-0_4

annotations he had made to Galen's anatomical writings, only a few of which he had introduced into his earlier revision of others' Galenic translations.[3] By mid-December 1543, at the latest, he was back in Padua, to the surprise of his deputy, Realdo Colombo, who had been boasting about his own prowess with the knife that had allowed him to reveal many errors of Vesalius.[4] Everyone wished to hear him, and over five hundred eager spectators came to watch him dissect a man and, most spectacular of all, a woman.[5] He continued a triumphant progress around Italy, dissecting at Bologna far into the night, before allegedly slipping away to Pisa, whose university had been reopened only a few weeks previously.[6] There he carried out a series of dissections on corpses gladly sent from Florence and elsewhere at the behest of Duke Cosimo de' Medici, who would himself have liked to attend at least some of the lectures. There were also autopsies in Pisa and Florence, and an attempt was made to persuade Charles V to let his physician remain at Pisa. The request was turned down.[7]

Vesalius' translation from professor to imperial courtier has often seemed strange to those for whom a major university chair might seem the summit of achievement. Yet family precedent, ties with his home region, wealth, honour, and the prestige of attending the greatest monarch in Europe will all have had their attractions.[8] He was also annoyed with the Venetians, who, angry at his dalliance with Pisa, withheld part of his salary.[9] His initial grumbles at the salary he received on his appointment on June 3, 1544, were soon modified by increases and further opportunities to gain additional wealth and honours, including a later appointment for his sister Anne as one of the *lavanderas de corpo*.[10] Nor should one forget that Vesalius was trained as a physician and that his interests also extended to therapeutics, one of the reasons why he returned to Venice for a final year.[11] Royal courts were also major centres of scientific culture and patronage; the physician-botanists Mattioli and Carolus Clusius or, among Vesalius' contemporaries, the prolific Galenist Giulio Alessandrino were courtiers as well as scientists.[12] Nonetheless, a decision to enter imperial service was, as Vesalius himself soon realised, a step that, once taken, was almost irrevocable. Despite the pleas of his friends, and as he came to regret later, he got rid of much of his library, including his unfinished writings, and his own copies of Galen, into whose margins he had written corrections and observations.[13] His scathing comments on the value of such marginalia will have also been directed to such opponents as John Caius, the margins of whose books are filled with copious annotations, and whom he wrongly believed to have been behind the London 'plagarism' in 1545 of the *Epitome* by his fellow Fleming, Thomas Geminus.[14] Becoming a court physician, Vesalius realised, meant the likely end to all his writing.[15]

Once he had settled his affairs in Italy, he returned North to campaign alongside the Emperor, treating the wounded, not always with total success, and preparing for the burial of the bodies of nobles and princes. Wherever Charles V went, Vesalius followed, to Ghent, to Brussels, to Cologne, to Nijmegen, to Ratisbon. He treated the Venetian ambassador, the envoy of the Duke of Ferrara, and other courtiers.[16] In 1544 he married Anne van Hamme, the daughter of a rich counsellor of Brussels, who bore him a daughter, also named Anne, in 1545.[17] It was around this time that he became fully aware of the furore that he had created among those whom he had often considered his friends, the Galenists. It was not unexpected, as his many disclaimers in 1543 show, and took various forms. By the end of the year, Niccolò Massa was being importuned by his students for his opinion of this "huge volume". In two letters sent to Professor Fracanzano at Padua at the end of January 1544, he confessed that he had not had time to read the whole book carefully, or even dip into it (a wilful exaggeration as his comments show). He agreed with many of Vesalius' observations, praising him for his learning and the obvious effort he had put into his many dissections. The students were delighted with the *Fabrica's* illustrations, even if they were not as accurate a representation as Massa had shown in his teaching, and there were errors in the description of the brain and the tongue. But its sheer size meant it was sometimes confusing to the students, although Massa admitted that he had been forced to omit many of his own discoveries from his shorter *Introduction* for reasons of space. He had two major objections: for all its length the book did not explain to the reader how the body worked, and, more seriously, Vesalius was ungrateful to his predecessors both in Antiquity and in the present. Many anatomists, Achillini, Berengario and himself, among them, had made discoveries that contradicted Galen and that were taken over unacknowledged by Vesalius. This ingratitude disturbed him: *ingrata nos turbant*.[18]

Massa, as an experienced dissector himself, took a more favourable view of what had been achieved than did Galenist translators and commentators. Vesalius' old housemate, John Caius, although relatively polite in public, filled the margins of his copy of the Basle 1538 Galen with criticisms of the errors and mistranslations of "Wesalius", many of them justified even if of little relevance to anatomy. A Paduan lecturer in 1550 bemoaned his spirit of contradiction and wished that he had been more modest in his criticisms of Galen.[19] Others were more forceful in their reactions. Janus Cornarius, who only a few weeks before the appearance of the *Fabrica* was sending him his best wishes, reacted with fury. He struck out all mention of Vesalius' name in his copy of the 1542 Basle edition of the Latin Galen with thick black ink, pressing down so fiercely onto the paper that the indentations can be still seen several pages later.[20] In Paris, his old teacher Sylvius delivered a series of caustic denunciations in front of his students that called forth a letter from Vesalius that formed the basis for the second part of his next book.[21] Printed by Oporinus in October 1546, it ostensibly discussed the so-called China root, a plant from the East that was said to be superior to guaiac as a remedy for gout, syphilis and a variety of other disorders but which, Vesalius averred, had been either neglected, misrepresented or improperly used[22] (Fig. 4.1). It is a competent little tract that shows Vesalius' wider interests as a physician, but it has never attracted the same interest as the revision of his letter to Sylvius that takes up almost three quarters of the book, and which restated succinctly and unequivocally the message of the *Fabrica*.[23]

It repeated what had been said there with greater force, although, if Dan Garrison is right, with less clarity in its Latin.[24] Vesalius insisted that there was no one more pious and deferential in his attitude to Galen than he himself; one could acknowledge Galen's skill as an animal dissector and as a propagandist for anatomy without the need to accept all that he had said.[25] Besides, the *Fabrica* was only a restatement of the more accurate human anatomy that Galen and his followers had obscured.[26] Human dissection was essential. The results of his own recent dissections, including those he had performed in Padua, Bologna and Pisa, only emphasised the continuous benefits that would accrue from plying the knife rather than engaging in wearisome disputes or pedantic marginalia. Sometimes dissections and, particularly, autopsies confirmed what he had written earlier, sometimes they added new material, for instance, about the internal organs of a woman, or revealed new anomalies.[27] But, far more than in the *Fabrica* itself, it is his growing divergence from Galen that stands out. Freed from the long expositions of the big book and underscored by the many headings in the margin, Vesalius' criticisms of Galen are sharper and the reasons for his disagreements more cogent. Allied to his comments about the demands of his new position as an

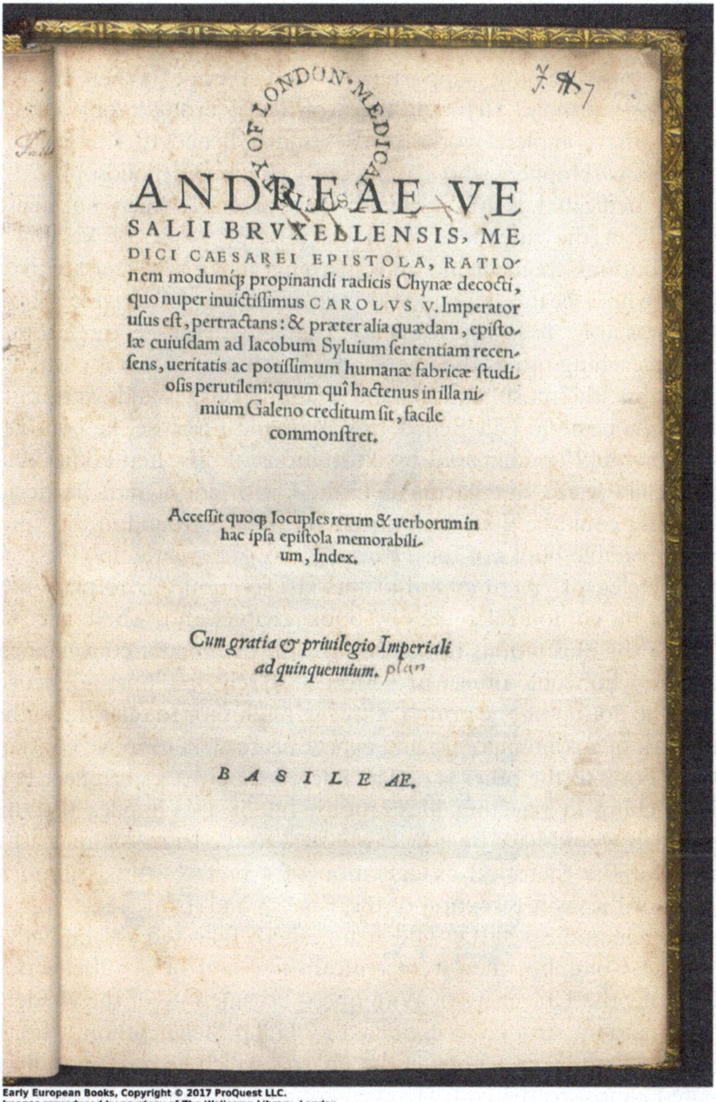
ANDREAE VE
SALII BRVXELLENSIS, ME
DICI CAESAREI EPISTOLA, RATIO-
nem modumq; propinandi radicis Chynæ decocti,
quo nuper inuictissimus CAROLVS V. Imperator
usus est, pertractans: & præter alia quædam, episto-
læ cuiusdam ad Iacobum Syluium sententiam recen-
sens, ueritatis ac potissimum humanæ fabricæ studi-
osis perutilem: quum quî hactenus in illa ni-
mium Galeno creditum sit, facile
commonstret.

Accessit quoq; locuples rerum & uerborum in
hac ipsa epistola memorabili-
um, Index.

*Cum gratia & priuilegio Imperiali
ad quinquennium.*

BASILEAE.

Fig. 4.1 Andreas Vesalius, *Epistola, rationem modumque propinandi radicis Chynae decocti….pertractans*. Basle, J. Oporinus, 1546, title page. (*Source* Wellcome Collection)

imperial physician, his vigorous defence of his position has an air of finality about it.

He was also gaining supporters around Europe. When in 1549, another Basle printer, Heinrich Peter or Petri, brought out a revised edition of the complete works of Alessandro Benedetti (d. 1512), its editor, Marcus Hopper, who taught both Greek and philosophy at the university, dedicated it to Vesalius.[28] His preface, which unusually is mentioned on the title page, contains fulsome praise of Vesalius for his contributions to medicine and anatomy and for his service to the Emperor, whose health depended on his expertise.[29] Everyone, Hopper averred, was now following Vesalius' example of dissecting for themselves and teaching others the importance of this skill. He was particularly impressed by "the most exquisite and beautiful images and expositions", and when he came to edit Benedetti's anatomical treatise, he considered that it too would be improved by illustrations of dissected bodies drawn as they really were. But having consulted with colleagues, he decided that nothing could be done to improve on what he found in the "divine and incomparable books of the *Fabrica*", and particularly upon its "consummately elegant" pictures and plates. His solution was to place in the margins of his edition references by book, chapter and, where necessary, number to the illustrations in the *Fabrica* for the greater convenience of his readers, who would otherwise be forced to rely on blunt and incorrect images. The notion of one printer allowing his author to refer throughout to the work of a contemporary and expect his readers to move constantly from one book to the other is remarkable. Not only does it reflect Petri's penny-pinching in rejecting illustrations, but it also implies that most professionals would have by now easy access to the large volume.[30] Even allowing for the rhetorical exaggeration of a preface, this is important evidence for the swift reception of the *Fabrica* and its message.

Younger anatomists in Italy and Spain eagerly followed Vesalius' advice, even if like Colombo, they were critical of several of his discoveries.[31] In Germany, the University of Wittenberg became one of the citadels of the new anatomy, strongly encouraged by Philip Melanchthon, who, like others there, had his own copy of the *Fabrica* which he read carefully and annotated almost to the end. He even composed a poem on it which he wrote into his copy. In 1550 he delivered a lecture on the importance of an education in anatomy and, two years later, aided by his colleagues Jakob Milich, Caspar Peucer and his Tübingen friend Leonhard Fuchs, he produced a second edition of his book *On the Soul* that replaced a

Galenist anatomy with one based on that of Vesalius. Not only did he praise him for his elucidations of Galenic terminology and his beautiful illustrations, but also cited him for his description of the epiglottis and the liver. Although he does not specifically mention Vesalius, he will have been among the "others" who are credited with rejecting the notion of a *rete mirabile* and of cotyledons in the human womb.[32] Melanchthon's revised book was influential in spreading the new anatomy in Lutheran Germany.

So also was Fuchs' *Epitome* of the *Fabrica*, which was published in two parts in 1551 and 1555.[33] It was intended to repair what he saw as a deep ignorance of anatomy and a lack of suitable instructors in anatomy in Germany. He rightly saw Vesalius as a continuator of Galen and his title proclaims a new synthesis of their anatomical writings.[34] Galen, he admitted, worked on animals, not on humans, and it was Vesalius who first described human anatomy carefully and correctly. Although he found fault with some of the latter's explanations, his main complaint was that neither the *Fabrica* nor the *Epitome* was suitable for beginners. One was too long and too detailed, the other too brief and requiring an understanding of some basic principles that could only be gained from the larger book. Fuchs rejected what he called the "insolence" of those who ranted immoderately against Vesalius, attacking him with lies and insults for demonstrating that the anatomy of Galen does not conform to the human body.

The most vociferous of these detractors was Sylvius, who denounced both Vesalius and Fuchs for their "insolent and ignorant slanders" against Galen and Hippocrates in an intemperate tract whose title declared it to be a refutation of "the calumnies of a certain madman against the anatomy of Hippocrates and Galen". (The Latin word for madman, *Vaesanus*, would instantly be interpreted by his potential readers as a pun on the name of Vesalius.) Not only was the *Fabrica* in his view a farrago of worthless information, as Fuchs would surely now realise, but it was far too long and filled with trivial and irrelevant statements. The illustrations, covered in shadows and a superfluity of letters, were more of a hindrance than a help.[35] It was a powerful denunciation of some of Vesalius' faults, but by this time his students, one of them reported, had grown weary of his implausible and offensive rhetoric.[36] Within little more than a decade after the publication of the *Fabrica* the tide had turned in favour of Vesalius. When in 1562 Francesco dal Pozzo delivered a long and vituperative attack on Vesalius, repeating Sylvius' arguments that Vesalius had misunderstood Galen's Greek, minimised Galen's experience of human bodies, and mistaken for normal anatomy what were clearly anomalies in the bodies had dissected, he was decisively rebuffed.[37] Gabriele Cuneo, one of the medical men he had wrongly invoked as an opponent of Vesalius, responded with an equally abusive and more successful refutation of dal Pozzo's bluster, and no similar denunciation of the major theses of the *Fabrica* was subsequently published.[38] Nonetheless, particularly in Spain under the leadership of Luis Mercado (1520–1606), there were still doubts about the value of Vesalian descriptive anatomy for physicians, as compared with Galen's functional anatomy of the body.[39] Knowing how the body worked and the purposes for which, it was believed, each part was created, as Galen had taught, was considered far more important for a physician whose job it was to restore a malfunctioning body to health than examining its structures in a corpse. That was why in Padua and elsewhere students continued to be referred to Galen rather than Vesalius for further elucidation of the structures of the body.[40] As late as the middle of the seventeenth century, at least for a while, candidates for a doctorate at Bologna had to swear an oath to preserve and defend the doctrines of "those who have been approved for many centuries now – namely Aristotle, Galen, Hippocrates, etc.….and not allow their principles and conclusions to be overturned or destroyed by anyone". In 1679, an anatomy lecturer there could end his course by advising physicians to abandon dissection as something useless, "performed only by persons of little talent and little brain", since the study of Galen's *On the Usefulness of Parts* was sufficient by itself.[41] Not all Vesalius' practical innovations were followed even in Padua and Bologna.[42] In Vesalius' absence in Basle in 1542, and again from 1544, the Paduan faculty reverted to the older method of a collaborative dissection, which was confirmed in the revised statutes of 1545.[43] Realdo Colombo, the professor of surgery, cut up the body, while the first ordinary professor of practical medicine, in 1544 Da Monte, who had just exchanged his chair in that subject for that of theory, explained what could be seen, while Crassus, the second professor, pointed to the relevant parts of the body.[44] But younger anatomists, as we shall see later, below, p. 189, came quickly to appreciate what Vesalius had done and set out to improve on, not abandon, his teaching.

Vesalius still maintained his anatomical interests, especially when performing autopsies, but his imperial duties accompanying the Emperor on his travels and attending to the ills of courtiers, ambassadors and the like took up his time. His patronage was sought even by a non-medic. Sigismund Gelenius (1497–1554), who had worked for Froben in Basle, dedicated his edition of the late-antique *Notitia Dignitatum*, a description of the duties of officials in the Late Roman Empire, to him in 1552.[45] Four years later, Vesalius was instrumental in obtaining from the Emperor a patent of nobility for the Basle printer Heinrich Petri, partly through gaining Petri's assurance that he would no longer publish Lutheran books.[46] His wealth continued to accumulate. He built a very large house in Brussels, with galleries, stables, and its own orchard, next to that of Count Mansfeld, and he was amassing a considerable fortune, made in part from his private practice in the city, where he may have resided almost continually from 1553 to 1556.[47]

But it was not long before he began contemplating a revised edition of both the *Fabrica* and its Latin *Epitome*. It is unclear to what extent, if at all, he had been involved in the issue by Oporinus around 1548–49 of a variant version of the former, slightly corrected and using in part a different font.[48] Vesalius himself never mentions it, and the small number of changes, mainly typographical, could have been made by the printer and his corrector without recourse to the author. The most interesting is the replacement of the initial letter at *5r with a slightly different one belonging to the same Holbein set as the initial that had been used earlier on p. 228/328.[49] Given that at least one other initial was showing signs of damage and that the replacement letter is only vaguely related to anatomy, Vesalius himself may not have been consulted.

For a very expensive volume the *Fabrica* had sold well: the copies that had been available in Leipzig had all been bought before the end of the year, when Joachim Camerarius could presume that his Erfurt friend Georg Sturz already had a copy. At Wittenberg, Melanchthon's encouragement, aided by some judiciously delivered free copies, ensured a steady sale in Lutheran Germany.[50] Copies were swiftly bought by several Oxford Colleges, and some Cambridge Colleges may have followed their example.[51] In, probably, 1545 a former Fellow of King's College, Cambridge, with an interest in medicine, Christopher Langton (1521–1578), published as the first part of a compendium of medicine in English 35 pages on the anatomy of the body along the lines of Benedetti, which he had judged as good as his "wit, learning, knowledge and the sterility and barrenness of the English tongue would allow".[52] Two years later, in 1547, Langton saw no reason to repeat his earlier detailed account of anatomy. Instead, he sent his readers off to consult for themselves the anatomical treatises of Galen and Vesalius, "for they have written whole books, and great volumes of them".[53] He was referring particularly to Galen's *Anatomical Procedures*, and *On the Usefulness of Parts*, and, presumably, to the *Fabrica* and, if more than one work by Vesalius is indicated, the *Epitome* or the *Institutions* also. This is, as far as I can tell, the first reference to the published *Fabrica* in a general account of medicine and, indeed, in any printed text not specifically concerned with Vesalian anatomy. But there is no indication that Langton had done more than glance at the big books of either Galen or Vesalius, and he displays no real interest in dissection other than what he had gained from books like those of Guinter and Laguna and perhaps Geminus.[54] It is, however, clear that, unless this is mere window-dressing, he expected at least some of his readers to be able to consult Vesalius' "great volume", which is further evidence for the spread of Vesalius' writings in England.

Thomas Geminus (fl. 1540–62), a Flemish engraver who had settled in London, was even quicker off the mark than Langton in promoting the Vesalian message.[55] He had begun planning a shorter anatomy in Latin within a few months of the publication of the *Fabrica* and its *Epitome*, for he had shown a draft, and, almost certainly, some of its illustrations for approval to none other than King Henry VIII in the early summer of 1544. His presentation was well-timed, for the king had recently approved the creation of lectures on anatomy, including demonstrations on corpses, for London surgeons, although his absence on campaign delayed the royal assent to publication until 1545. Geminus claimed to have taken the copper plates in this Latin *Compendiosa totius anatomiae delineatio (A Comprehensive Sketch of the Whole of Anatomy)* from the *Epitome*, which Geminus thought too brief, although far cheaper and less prolix than the *Fabrica* itself.[56] He certainly took his male and female figures from the *Epitome*, but most were selected from the larger work. Vesalius thought little of the result, and one need not take seriously his comment that, had he been informed of Geminus' plans in advance, he would have been happy to lend him the original wood blocks, whose quality, he declared, was far superior to those Geminus had used.[57] Others were more grateful. Geminus' illustrations were, in turn, copied in many similar publications such as the 1551 *Anatomia Deutsch*, a new German translation of the *Epitome* for surgeons, and in Nicholas Udall's partial English version that appeared in 1553.[58]

The 1555 Edition

Given Vesalius' habit of correcting and adding new material to his publications almost from their moment of issue, as we have seen with the *Paraphrase* and the *Principles of Anatomy*, one might assume that he would have done the same with his first edition of the *Fabrica*.[59] But turning any marginal corrections into a new edition of a substantial and illustrated book was a different matter entirely. Besides, unlike the revisions of the *Paraphrase* and the *Principles of Anatomy*, which were short university textbooks aimed at a wide audience around Europe that was constantly renewing itself, the *Fabrica* was large, complex and expensive. Nor could it be expected that all its original purchasers would buy it again. Vesalius' self-confidence and Oporinus' ambition had won the day in 1543. That they would do so again required a good deal of optimism.

It is possible that Vesalius first told Oporinus that he was planning a new edition in a brief visit to Basle in 1547, or a year or soon after. He borrowed the matrices from which the fonts for his new edition were cast in late 1549 or a little later, and printing may have begun soon after.[60] Certainly, the work was sufficiently far enough advanced by December 1551 for Oporinus to tell a friend that a new edition would soon appear.[61] Five months later, in May 1552, his sales catalogue offered for sale the first five books, revised and enlarged, as well as copies of the German edition of the *Epitome*, and Conrad Gessner could optimistically include a new edition of 1554 in the additional volume of his bibliographical masterpiece, his *Bibliotheca Universalis*, *A Universal Library*, that was due to appear on early 1555.[62] But two things were needed before the revised *Fabrica* could do so, money and materials, and both were lacking.

The considerable expense involved in the initial processes of producing any book meant that all publishers constantly teetered on the edge of bankruptcy. Oporinus was no exception, for by mid-1552, he was facing a financial crisis. Superstitious friends blamed his inauspicious choice of Arion and the dolphin for the *Fabrica*'s colophon. They questioned the suitability of a musician who was rescued by a dolphin after diving into the sea to escape being killed by pirates and who then found refuge at the court of a tyrant. A more likely cause was the problems he was having with the authorities and in disposing of his stock.[63] Selling only part of the revised *Fabrica* would ease his cash flow, but, given the amount of cross-referencing in the *Fabrica*, this would be unsatisfactory until the final pagination was known for the whole work, even if buyers were prepared to wait for the volumes until they could purchase the last two books as well.[64]

His financial worries were eased by a huge loan from Vesalius.[65] How large is unclear, but it allowed printing to go ahead on terms that were unfavourable to Oporinus. When in 1567 Oporinus sold his business, he made special arrangements for the sale of two items, the *Fabrica* and the equally remarkable monumental genealogical tables of the rulers of the Palatinate, Johannes Basilius Herold's *Tabula palatinorum*, that he had printed in 1556. Income from their sale was to go directly to their authors, presumably to pay off loans.[66] In 1569, following Oporinus' death the previous year, Vesalius' descendants engaged in a lawsuit to secure the return of the wood blocks for the *Fabrica*, the copy used by the press, any unsold copies of his books and the considerable sum of 2582 guilders that they claimed they were still owed. This sum was larger than almost any other repayment made by Oporinus and may well have been intended to cover most of the costs of publishing the second edition of the *Fabrica* in 1555. Whether they were successful is not recorded, but the debt still had not been settled almost a decade later, when Henrich Vandermeeren, who had married Vesalius' widow, was still trying to recover it.[67]

The delay caused an unexpected problem. When in May 1555 it was decided to go ahead with the missing books, Oporinus wished to print them in exactly the same style as the earlier ones with the same font and the same size of paper. Neither was immediately to hand in Basle. He wrote several letters to his Strasbourg friend, the preacher Conrad Hubertus, asking for his help to buy sixty or seventy sheets of paper of the same size as that in the first edition. But appropriate paper was easier to find than appropriate fonts. As he wrote to Hubertus, the original type that had been used in 1551 had become, or was likely to become, badly worn and needed replacing, and, worse, he no longer had access to the original matrices from which it had been cast.[68] He had borrowed them in 1549–50 from a Spanish evangelical humanist, Francisco de Enzinas (1518–1552), who was revisiting Basle in the course of his wanderings to find a suitable printer for his writings and Spanish translations. He and Oporinus had once been friends. In 1546 Oporinus had published his *Historia vera, A True History*, which described the life and murder of his fellow Spanish religious exile, Juan Diaz, and became a European bestseller.[69] The following year he printed at Enzinas' expense the much larger *Historia imperial y cesarea, A History of the King and Emperor*, of Pedro de Mexia, as well as his Spanish translation of some biographies by Plutarch. Enzinas then moved to newly Protestant Cambridge at the urging of Martin Bucer, where he taught Greek briefly from 1548 to 1549. He next transferred his young family to Strasbourg before travelling around Switzerland in search of a publisher for his Spanish translations and, not least, for his Spanish translation of the Bible. He found one in Zurich, the impecunious Augustin Fries, with whom he planned to publish more versions of Plutarch's writings, before returning to Basle at the end of 1549. His relations with his old publisher were at first cordial. He allowed Oporinus to borrow the matrices he had planned to use in his edition of the Bible in order to cast the elegant letters for his new edition of Vesalius, but soon everything changed. He had hoped to bring Fries with him to set up in business as a printer of Spanish books, but this attempt to break into the Basle cartel provoked angry responses from printers and from clerics, who suspected him of being a spy. Oporinus broke off all contact with him in spring, 1550, and the town council passed a law forbidding the printing of any books that were not in Latin, Greek, Hebrew or German.[70] Enzinas was forced to move back to Strasbourg, where he found a more congenial hearing, remaining there until his death from plague at the end of December 1552.

His death, which would have been known in Basle, did not trouble Oporinus at first, for he stated on more than one occasion that he wished to have nothing to do with him. But by May1555, he was becoming desperate to locate the matrices he had used earlier, for without them, the remaining books might have had to be printed in a visibly different font. That was why he asked Hubertus and, possibly, other friends to help find them. The search was evidently a swift success—they may have still been in the possession of Enzinas' widow or a Strasbourg printer—, for the colophon to this edition is dated to August 1555, although publication may have been postponed for a few weeks in order to receive and print Vesalius' corrections to the proofs that were subsequently included in the *Errata*.[71]

As well as being considerably longer than its predecessor, 824 pages compared with 663, the new edition presents many subtle differences in its appearance. Although the page size is effectively the same, the relationship between the roman and italic fonts is changed. The new roman font is larger than before and with fewer characters per line, while the italic is now smaller, with more characters per line. The italic font, which presents what Roger Gaskell calls a dazzling appearance on the page, is also narrower than its predecessor. Whereas in the first edition the two fonts were in balance and equally legible, the continuous text is now easier to read and dominates.[72] Italic instead of roman is used in the headlines, and there is slightly more space than before between the letters of roman type and slightly less between the italic. The actual text block in 1555 is smaller, leaving more white space surrounding the text, as Vesalius had demanded of Oporinus in setting his *China Root Letter*. Where the illustrations to the text were allowed to remain unchanged, the cutter was asked to make some of the identifying letters in them clearer by removing a tiny amount of the wood block, not always to Vesalius' satisfaction.[73]

The difference in the type size required an almost total redrawing of the chapter initials. For the larger initials Oporinus, where possible, utilised the original blocks, but a new block had to be cut for the letter V, because an original *QVANTVMVIS* (sig. a*2) and *QVANDOQUIDEM (*384*)* had been replaced by *VTCVNQVE* (a* 2r, 589). This new letter is unique in that it represents a new theme, the flaying of Marsyas while two women look on, a classicising insertion that is out of place in both style and content (Fig. 4.2). Barry Anson, who was one of the first to devote attention to the 1555 chapter initials, noted that one of the representations of the letter L was printed from an original block, but without investigating further.[74] This is the only chapter heading within the book that was not changed, and, significantly, it was not printed from a block cut from the designs Vesalius had sent from Padua in 1542, but, as has already been discussed, it is one of a series of Holbein letters and introduced later in Basle.[75] This slightly larger image, in turn, formed the model for the size of the new letters.

These are based on the 1543 initials, but in a very different style and cut more sharply to give a clearer image, as in letter C.[76] Oporinus made a definite effort to repeat the same themes as in 1543, albeit with some significant differences. The size of the blocks is a little larger than before, about 4.1 cm compared with 3.7 cm, and the figures are now smaller, allowing for more background to appear. In the letter O the man in the hat is now inside the letter itself, so creating space behind him, while in the standard letter L the fact that this is a criminal being taken down from a gallows is emphasised by the presence of a hooded figure of a monk and a cross in the background. In the M the smaller figures are placed in front of a distant landscape with a church spire. As in the C and the F many of the human figures are replaced by *putti* performing the same actions, and the number of individuals is frequently increased, usually by one, as in the

Fig. 4.2 Andreas Vesalius, *De humani corporis fabrica*, Basle: J. Oporinus, 1555, p. i. The new opening initial uniquely depicts a classical scene, the flaying of Marsyas. (Copyright: Christie's)

P, S and T. In the letter V the three figures of 1543 become four *putti* in 1555. Several of the designs were clearly based on the printed edition of 1543, although in some the arrangement of figures is now reversed.[77] The opening letters in the index are also changed. For the opening A Oporinus used the same initial from the text as in 1543, but only five other initial letters, F, G, O, S and Z are identical in both editions. In 1543 only one index heading, for F, shows a *putto* defecating or showing his backside, whereas in 1555 the same or a similar image is used also for E, L and T. The other initials are very varied and were probably put together from what Oporinus had to hand.

These changes in the initials show how much Oporinus was determined to keep up the high standard of workmanship of the first edition. Their superiority to those of 1543, both in the organisation of the design and in the clarity of the printed image, is not necessarily to be credited entirely to the superior skill and draughtsmanship of the artist and the cutter, presumably both working in Basle. This type of initial showing several figures within the block was rare until the late 1530s, and, as in so much else, Vesalius and Oporinus were pioneers. Fifteen years or so later, such initials had become much more common, and those involved in producing them had become more experienced.

The quality of these initials contrasts with that of the revised title page, which may have had to be recut because of damage to the wood block (Fig. 4.3). Although based on its predecessor, it shows slight differences in perspective as well as a somewhat coarser draughtsmanship or cutting. Vesalius' head, already slightly large in 1543, is now drawn even larger to become almost a caricature. Some changes are easily explicable; Oporinus' possible monogram on the column at the top has now gone, replaced by his name as publisher, which is now set in type at the foot of the page in large capitals. The cartouche at the bottom of the page with the imperial privilege is now drawn to represent the board on which animals were tied for dissection, and the wording has been reset. The poses of many of the audience, as well as that of the skeleton (who now carries a scythe) are subtly changed. Most striking of all, the nude figure holding on to the pillar on the left is now elegantly clothed in tunic and cloak. The classical figures are also dressed slightly differently than before: Galen's outer garment has been shortened and has lost its elegant fringe, while his classical sandals have become slippers. The weasels in Vesalius' coat of arms now appear more weasel-like, and he has exchanged his title of professor for that of physician to the "supremely undefeated Charles V". Compared with the revised initials, the overall clarity of the 1543 title page is diminished.

By contrast, the text of the second edition is much improved.[78] There are more marginal notes and the references are more precise, although the list of errata is much longer than in 1543. Vesalius constantly altered individual words sometimes to give greater precision, sometimes on grounds of style, abolishing wherever possible the old Arabist anatomical terminology.[79] Some changes remove what a dozen years later now appeared inappropriate or irrelevant, such as the marginal comment on the *glossokomion*, although the initials deriving from it still remain in altered form for E and F.[80] Vesalius omitted the names of those who helped him but who had either died in the interim, such as Gerard van Veltwyck and Giambattista Da Monte, or, like Sylvius, Colombo and Fuchs, were friends no longer. The way in which both Charles V and the future Philip II are now mentioned in the preface may indicate, as O'Malley suggested,

Fig. 4.3 Andreas Vesalius, *De humani corporis fabrica*, Basle: J. Oporinus, 1555, title page. (Copyright: Christie's)

that Vesalius was aware of Charles' decision to abdicate as Emperor, although he had not yet done so when the book was published.[81] Vesalius no longer discusses the shapes of skulls and those found in Alpine cemeteries or explains in detail how a corpse should be suspended for further study.[82] Other passages and even chapters are moved about. In Book I, the original order of Chapters XXXIX and XL is reversed and a new chapter is added by transferring Book II, Chapter VII, to follow them, with a consequent renumbering of the book's later chapters. Chapter II of Book II is drastically shortened, while a passage on the ligament of the tongue in Chapter XX of that book is moved to the end of Chapter XI.

Some changes may reflect Vesalius' awareness of a possibly growing dislike of anatomical dissection among the general public as something cruel and close to illegal in its disregard for propriety. The actions of students (including the young Vesalius) in rifling tombs and cemeteries for bones no longer appear.[83] The story of the students' theft of the corpse of a monk's mistress remains, but the protesters against Vesalius are reduced from the monk and his mistress's kin to the monk and her mother, one a sinner, the other intellectually and morally weak.[84] His potentially shocking account of investigating the still beating hearts of two individuals who had just died is toned down by saying that one had died in an accident, while the other had been executed.[85] Theological considerations may also have encouraged him to omit some of his comments about the soul, as well as his passing reference to the blood and water dropping from the side of the crucified Christ, for the pressure of the Inquisition was growing heavier.[86]

Other alterations reflect his responses to criticism as well as the development of his own ideas, some already adumbrated in the *China Root Letter*. His rejection of the Galenic interventricular septum in the heart is now total. In his first discussion, he had declared that, although the pits in the septum might suggest, as they had done to Galen, that some blood might be able to escape through them from one ventricle to the other, he had never observed that the septum was pervious. He reiterated this in 1555, revising at the same time his account of the vena cava and insisting that he was far from satisfied even with his own conclusions, although he now doubted ever more strongly the typical conclusions of his fellow anatomists.[87] His experiences with pregnant women allowed him to revise his ideas about the coverings of the foetus, distinguishing the shape of the placenta in women from that in dogs and that in ruminants, which alone had true cotyledons.[88] He illustrated other findings with a recut drawing of the laryngeal cartilage, as he did to illustrate the ginglymus hinge (for which he had been criticised by Caius), and the aorta.[89] But co-ordination between artist and author was not as good as in 1543, for Vesalius was not in Basle to see everything through the press. For example, the illustration of the five branches of the portal vein is retained, although the description is omitted, and, conversely, the missing branches of the nerve that Vesalius himself had admitted in 1543 had been left out through his oversight were still not replaced in 1555.[90]

Vesalius' policy of rarely mentioning his contemporaries makes it difficult to determine the source of any new information contained in the volume. His correct depiction of the ginglymus hinge he almost certainly took from John Caius, and he may well have been encouraged to revise some of what he had said about the valves in the veins by his contacts with Canani or from reading Amatus Lusitanus.[91] He knew of the detailed criticisms of his work by Sylvius, and probably by others such as Massa, but there is no clear evidence for his use of Estienne. He added a reference in his preface to his dissection at Pisa early in 1544, and in his improved account of the venous valves he refers to comments made by his audience at another dissection and to the belief, often repeated during his university anatomies, that the venous walls were made thick in order to provide strength.[92] His careful formulation of his reasons for rejecting Galen's interventricular septum makes it impossible to say if this was the result of more recent dissections or of meditating further on what he had already seen before 1543.[93] There are signs that he had been able to carry out some animal dissection as well as an occasional autopsy or clinical observation, but the demands of his imperial master and of patients were an obstacle to a systematic revision or expansion of what he had said before.[94] He was now more confident of the need to move away from following the letter of Galen's law and of the importance of dissection as a tool for understanding the body but his extensive alterations to his text and images improve the precision and accuracy of what he had already achieved in 1543 rather than contribute new information.

The fortunate survival of a single copy of the *Epitome*, dated 1555, now in the Waller Collection in the Uppsala University Library, classmark Waller Rar. St. fol., 9909, indicates that this too was reissued then by Oporinus.[95] However, the only change in it would appear to be the new final page, printed on a differently watermarked paper from that of 1543 and identical to that used for the 1555 *Fabrica* and with a new date for the colophon of April 1555. Such a substitution of a newly dated page was not uncommon, especially if, as happened with the later Juntine editions of the Latin Galen, the earlier printing had not sold out, and this may well have been an initiative by Oporinus without involving Vesalius. But Thorer's German version was not reissued again, and Vesalius may have lost interest in these student summaries.

The Last Years

The events of the last decade of the life of Vesalius are largely unknown. His one publication, a letter in response to Falloppia's criticisms, was sent to press only a few weeks before his death in 1564, although it may have been drafted as early as December 1561 and may not have been intended to be printed. He remained in court service, attending, albeit unsuccessfully, distinguished patients such as Ferrante Gonzaga (1507–1557), one of the military commanders of the imperial army, who died after being wounded at St Quentin in 1557, and Henri II of France, who died after a jousting accident in early July 1559.[96] Shortly after this, Vesalius moved with his wife from Brussels to Spain, with special responsibility for the health of the non-Spanish courtiers. He found it an even less favourable environment for his anatomical interests. Younger Spanish anatomists were keen adherents of his new anatomy, but he seems not to have related to them, complaining that, compared with the happy life that he had once enjoyed at Padua, he now saw little possibility of performing dissection and indeed had difficulty getting hold of a skull.[97]

The fortunate identification of the copy of the 1555 edition owned and annotated by Vesalius himself not only extends the history of the development of the *Fabrica* for a few more years but provides us with another glimpse of Vesalius at work that is more extensive and informative than the similar annotations to the *Principles of Anatomy* or the marginal additions to the second edition of the *Paraphrase*.[98] It allows us to see into Vesalius' mind as he considered and reconsidered how best to express his ideas, as well as illustrating his gifts as an artist who was also acquainted with the methods of printers (Fig. 4.4). It offers a unique insight into the mind of one of the great names in Renaissance science and medicine, differing from the reports of his and others' lectures in showing his hesitations, second and even third thoughts. Problems remain, most notably in the date or dates at which Vesalius made his notes, but their value is unquestionable.[99]

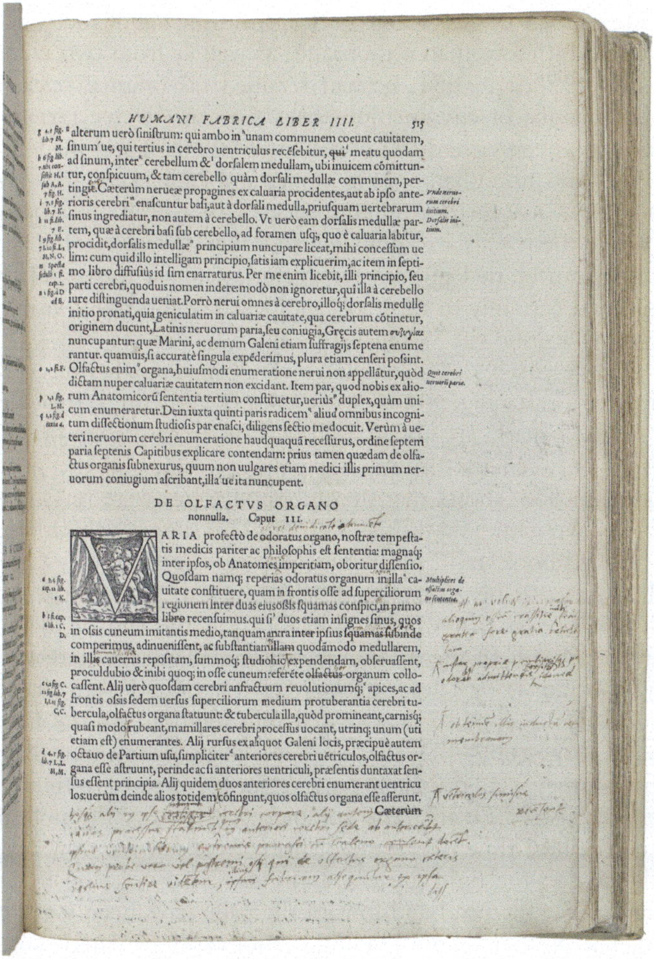

Fig. 4.4 Andreas Vesalius, *De humani corporis fabrica*, Basle: J. Oporinus, 1555, p. 515. Vesalius rewrites the opening of the chapter on the organ of smell. (Copyright: Christie's)

That the annotator was Vesalius is clear from the abundance of corrections, numbering more than a thousand, as well as from two comments that he made.[100] At p. 654, he had retained his original reference to the Jewish practice of circumcision that "we can still see (*cernimus*)", but in his note, he relegated this possibility to the past and made it more specific: "as I saw more than once in Venice and Padua (*non semel Venetiis et Patauii vidi)*". A further reminiscence of Northern Italy occurs on p. 721, where the tonsils are compared to stakes driven into the shore to which a boat can be tied up, a comparison unfamiliar to the inhabitants of Brussels and certainly to those of landlocked Madrid, but visible daily to those living around the Venetian lagoon.

The volume itself is incomplete. The final gatherings, Bb-Ee, containing the Errata page, the index, the list of gatherings, the colophon and the printer's mark, are absent, and may never have been present, for the contrast between the grubby last page, p. 824, and the preceding near-pristine p. 823, shows that the unbound pages were for some time placed face down gathering dust and grime. This suggests that these were not the finished book itself, but the incomplete sheets sent by Oporinus, as was his wont, to his distant author for him to read before sending back a list of *Errata* that would be included at the end of the volume when it was printed. In the meantime, Oporinus and his staff would be preparing the index.[101] These sheets were not bound immediately but used by Vesalius as copy for a potential third edition. By the time they were bound, pages 669–672, the inner leaf of gathering mm, had been lost, although they were originally present, for corrections to them appear in the printed list of *Errata*. The volume itself has been bound at least once after Vesalius' time, for many of his corrections are now deep and almost illegible in the tight gutters, and letters and occasionally lines of the larger ones at the head and foot of the page have been lost through cropping. The present vellum binding cannot easily be dated, but Vesalius' name is printed on the spine as A. WESALE, which might indicate that it had long remained in modern Belgium, perhaps in an institution rather than in private possession.[102]

Although many of the underlinings indicate private reading, Vesalius had intended his corrections to form the basis for a new edition, indicating changes with signs for the printer and specifying where exactly he wished things to be placed. On five occasions, he rewrote in the blank space within an illustration individual letters that had become obscured through the wearing away of the blocks, showing by arrows where they should go. On p. 241, he wanted three letters, B, C and D, replaced to make clear the triangle of musculature mentioned in the caption, while on p. 246 and in the text on p. 247, he changed the Greek compendium for ο and υ to the single letter N as demanded by the sense and adjusted the caption to fit. He told the printer how to do this: "a sliver of wood with the correct N should replace the ου".[103] Occasionally the block-cutter, in attempting to gain a clearer space around each letter within an image, had cut too deeply, removing part of the hatching of a tendon on p 232 and part of the outline of the toe on p. 234 (Fig. 4.5). Both here and possibly on p. 259 Vesalius redrew the missing lines with a very fine pen. One other correction should be noted as it represents the only correction in an illustration that depends on the change of mind on the part of the author. On p. 182, following the renewed discussion of what Vesalius (wrongly) saw as a cartilage rather than the muscle lifting the eyelid, Vesalius redrew its outline in the shape of a bell curve.[104]

Alterations to the captions themselves are much more extensive (Fig. 4.6). Some go back to the words or sentences that the setter had overlooked when he set the pages in a more spacious manner, as on p.109, line 6, and some correct even earlier mistakes that had defied his own and the printer's eye at least twice, such as the misplacement on p. 155 of the captions to KK and LL. On p. 245, lines 18–19 he changed the phrase "*et characteres...obuii*", reverting to wording akin to that in the 1543 edition, p. 155. He occasionally added new material. At the bottom of

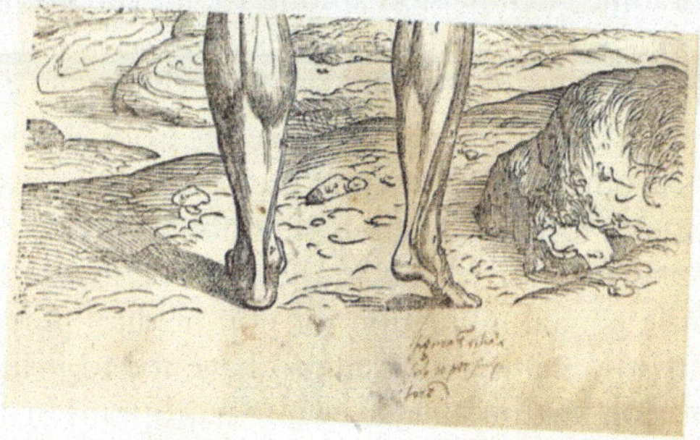

Fig. 4.5 Andreas Vesalius, *De humani corporis fabrica*, Basle: J. Oporinus, 1555, p. 254. Vesalius redraws an outline badly cut in 1543. (Copyright: Christie's)

p. 207 Vesalius wrote in the margin that the Roman author Celsus had used the word *brachium* to mean the forearm "contrary to all medical custom" (a note worded slightly differently in the 1555 Errata) and, a little above, the Greek term for ribs, σπάθαι.[105]

Although the great majority of the changes are stylistic, some were derived from his reading or, more rarely, from his knowledge of what was going on elsewhere. Often, he made his references to the ancient authorities, Hippocrates, Celsus, Galen, Oribasius and Paul, and even Cicero, more specific and more precise. By contrast, what he took from his contemporaries is disguised in silence or anonymity.[106] The reader was unlikely to know the source of the alternative term for the vertebrae, *verticula*, or to divine who it was who had been so ridiculously concerned about the sexual activity of pregnant women that they imagined channels bringing the seed direct from the ovaries to the neck of the uterus rather than to its cusps.[107] Those who oppose his view of the location of the organ of smell are divided into two camps, but his description of the opinions of the first group shifts from the pejorative "they imagine" (*confingunt*) through the more positive "they declare" (*statuunt*) to, finally, the neutral "they reckon" (*recensent*). His own superiority is enhanced by emphasising that a wrong view of the tonsils is now "widespread", and the addition of the modest "as far as I know" does little to modify his new assertion that his discoveries about the glandules around the larynx have been neglected by other writers. Indeed, it only intensifies the impression of Vesalius' scrupulous learning.[108]

The most striking of such additions comes late in the book, when, after discussing Pharaonic circumcision among the Egyptians, he refers to the custom of both male and female circumcision "continuing today" in Ethiopia among the Scaevani for religious reasons[109] (Fig. 4.7). These subjects of Prester John cut off the fleshy processes from newborn girls in accordance with their religion in the same way as they remove the foreskins of boys, "although in their religious ceremonies they are otherwise similar to those of us Christians". This is arguably the first reference in a medical text to female genital mutilation for non-medical purposes, since the context shows that Vesalius is unlikely to have found it in a medical source. This information first became known to Europeans in 1540 when the Portuguese humanist Damian de Goes mentioned it in a controversial treatise on the faith, morals and religion of the Ethiopians.[110] He did not, however, name the Scaevani there, but his collaborator, Paulo Giovio, did

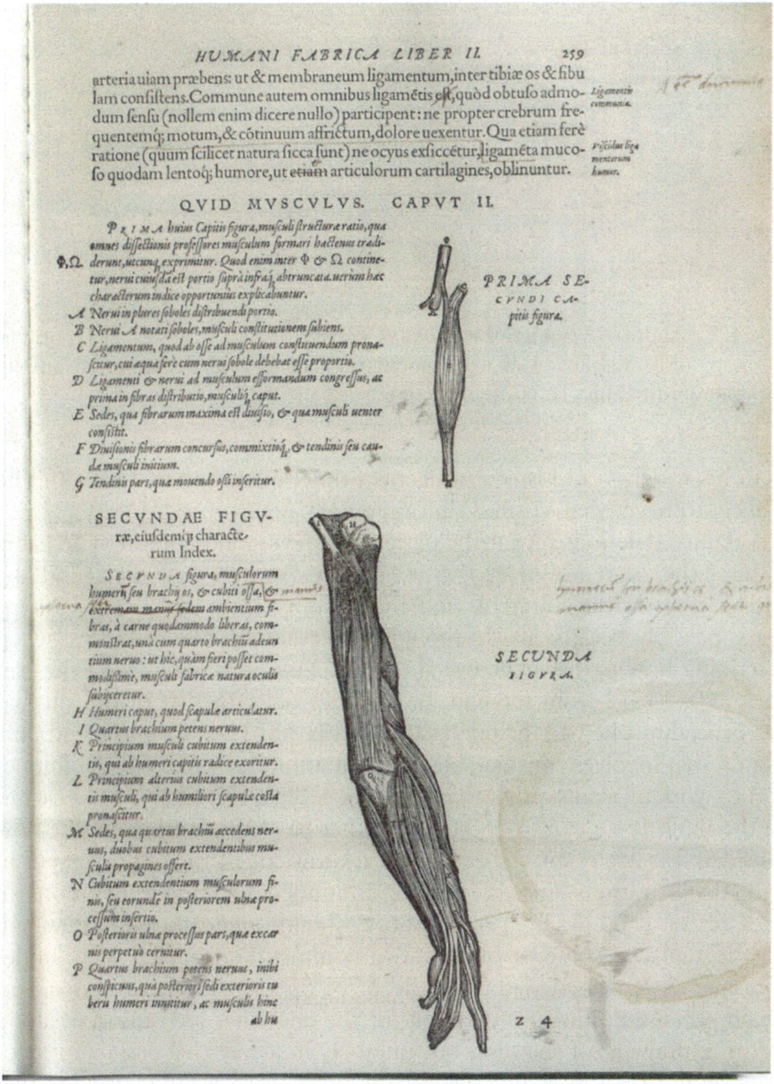

HVMANI FABRICAE LIBER II. 259

arteria uiam præbens: ut & membraneum ligamentum, inter tibiæ os & fibulam conſiſtens. Commune autem omnibus ligamentis est, quòd obtuſo admodum ſenſu (nollem enim dicere nullo) participent: ne propter crebrum frequentemq; motum, & cōtinuum affrictum, dolore uexentur. Qua etiam ferè ratione (quum ſcilicet natura ſicca ſunt) ne ocyus exſiccentur, ligamenta mucoſo quodam lentoq; humore, ut etiam articulorum cartilagines, oblinuntur.

Ligamentis communia.

Viſcidus ligamentorum humor.

QVID MVSCVLVS. CAPVT II.

PRIMA huius Capitis figura, muſculi ſtructuræ ratio, quã omnes diſſectionis profeſſores muſculum formari hactenus tradiderunt, utcunq; exprimitur. Quod enim inter Φ & Ω continetur, nerui cuiuſdã eſt portio ſuprà infràq; abtruncata. uerùm hæc characterum indice opportunius explicabuntur.

Φ, Ω

A Nerui in plures ſoboles diſtribuendi portio.

B Nerui A notati ſoboles, muſculi conſtitutionem ſubiens.

C Ligamentum, quod ab oſſe ad muſculum conſtituendum pronaſcitur, cui æqua ferè cum nerui ſobole debebat eſſe proportio.

D Ligamenti & nerui ad muſculum efformandum congreſſus, ac prima in fibras diſtributio, muſculiq; caput.

E Sedes, qua fibrarum maxima eſt diuiſio, & qua muſculi uenter conſiſtit.

F Diuiſionis fibrarum concurſus, commixtioq;, & tendinis ſeu caudæ muſculi initium.

G Tendinis pars, quæ mouendo oſſi inſeritur.

PRIMA SECVNDI CAPITIS FIGVRA.

SECVNDAE FIGVRÆ, eiuſdemq; characterum Index.

SECVNDA figura, muſculorum humerū ſeu brachij os, & cubiti oſſa, & extremam manus ſedem ambientium fibras, à carne quodammodo liberas, commonſtrat, unà cum quarto brachium adeuntium neruo: ut hic, quàm fieri poſſet commodiſſimè, muſculi fabricæ natura oculis ſubijceretur.

H Humeri caput, quod ſcapulæ articulatur.

I Quartus brachium petens neruus.

K Principium muſculi cubitum extendentis, qui ab humeri capitis radice exoritur.

L Principium alterius cubitum extendentis muſculi, qui ab humiliori ſcapulæ coſta promaſcitur.

M Sedes, qua quartus brachiū accedens neruus, duobus cubitum extendentibus muſculis propagines offert.

N Cubitum extendentium muſculorum finis, ſeu eorundē in poſteriorem ulnæ proceſſum inſertio.

O Poſterioris ulnæ proceſſus pars, quæ excarnis perpetuò cernitur.

P Quartus brachium petens neruus, inibi conſpicuus, quà poſteriori ſedi exterioris tuberis humeri innititur, ac muſculis hinc ab hu

SECVNDA FIGVRA.

Z 4

Fig. 4.6 Andreas Vesalius, *De humani corporis fabrica*, Basle: J. Oporinus, 1555, p. 259. Vesalius rewrites a caption about the muscles. (Copyright: Christie's)

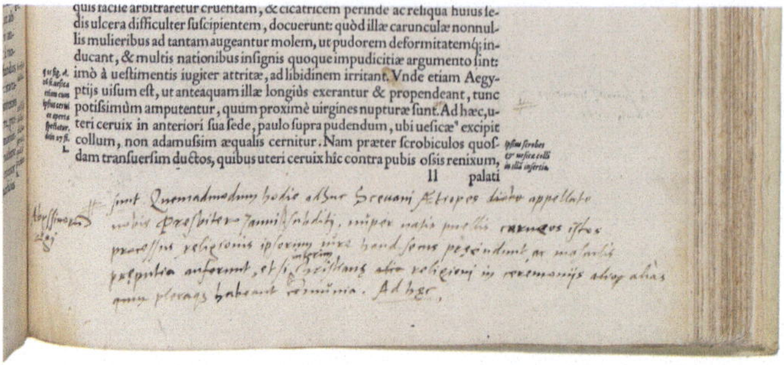
quis facile arbitraretur cruentam, & cicatricem perinde ac reliqua huius le-
dis ulcera difficulter suscipientem, docuerunt: quòd illæ carunculæ nonnul-
lis mulieribus ad tantam augeantur molem, ut pudorem deformitatemq; in-
ducant, & multis nationibus insignis quoque impudicitiæ argumento sint:
imò à uestimentis iugiter attritæ, ad libidinem irritant. Vnde etiam Aegy-
ptijs uisum est, ut anteaquam illæ longiùs exerantur & propendeant, tunc
potissimùm amputentur, quum proximè uirgines nupturæ sunt. Ad hæc, u-
teri ceruix in anteriori sua sede, paulo supra pudendum, ubi uesicæ excipit
collum, non adamussim æqualis cernitur. Nam præter scrobiculos quos-
dam transuersim ductos, quibus uteri ceruix hîc contra pubis ossis renixum,
Ll palati

Fig. 4.7 Andreas Vesalius, *De humani corporis fabrica*, Basle: J. Oporinus, 1555, p. 653. Vesalius adds a comment on female genital mutilation among the Scaevani in Ethiopia. (Copyright: Christie's)

so in his widely read history of his own time, published in 1550, and he is likely to have been Vesalius' informant.[111]

Vesalius' deletions repeat the pattern of the second edition. The name of his father is removed at f. A5r,22 since Andreas had died, and the information on the Ligurians and their tombs on p. 111, 25–31 disappears, possibly because it was thought irrelevant. In two instances, when describing how to dissect an eye at p. 287, 24–6, and a muscle of the arm at p. 390, 14–12 from the end, he includes only the first of the alternative procedures he had included earlier. At p. 68, 19–9 from the end, in a section that involves considerable reorganisation, he leaves out the injunction to work on skulls rather than pictures because the sutures were clearer on the skull. It is open to question whether he omitted it because he now considered it irrelevant or because he acknowledged the improvement in printed illustrations since 1543. He continued to remove mentions of the Creator at pp. 5, 17–19 and 366, 3–7, although others are retained, and he replaced some of his negative comments on Galen with kinder phrasing. Seven lines in the middle of p. 55 on the *discus articularis*, which acts as additional protection for the jawbones are removed, along with the drawing of the disc, but uniquely in a way that results in incoherence and the disc is not referred to again. Had Vesalius decided that this was, as he said, something that was common to all joints, or rejected what he described as an occasional anomaly? At pp. 571–2, his long revision of what he had said about the glands in the throat is a masterpiece of evasion. In 1543, p. 579, he had remarked that he refused to use the

Greek word *paristhmia* for glands at the base of the larynx because they had not been described in this way by professors of anatomy, implying that they were his rediscovery. The words remain in the second edition, but the annotation calls them "ancient professors", thus strengthening his claim to novelty. A similar claim for primacy in the discovery of the epispastic ligament in the larynx remains in the description of the larynx at pp. 308–9 and again on p. 720, but it is quietly abandoned at the first mention of the ligament on p. 274, 1, where the words "discovered by us" are deleted. In many passages, Vesalius' original statements are qualified by words such as "in some way", "perhaps" or "almost".[112]

Significantly in view of his time spent with the Emperor after 1555, nowhere in his annotations does Vesalius make any overt claim to a discovery, nor is there any clear evidence that in this period he had dissected a corpse. The overwhelming majority of changes that refer to dissected material are the result of meditating upon what had been done before rather than any new dissection. Several reflect his awareness of the difference between what can be seen in a corpse and what is now visible on the printed page: Words like 'downwards' or 'outer' are replaced by 'backwards' and 'frontwards'. Choosing the right word to describe the size or texture of a bone or organ is a major concern, as it had been in 1555. But this need not imply that Vesalius had carried out any recent work on a corpse, although the fact that most of these corrections occur in the first two books, dealing with bones, muscles and ligaments suggests that he might have had some occasional access to skeletal material.

Only in three places does Vesalius add new anatomical material. In his chapter on the foetus, which had been extensively revised in 1555, he remarks on p. 481 that the foetal sac is attached more strongly to the rear of the uterus because the vascular network is denser there and a special receptacle for the urine of the foetus is located between two of the wrappings of the womb. The second occurs in the heavily altered section on the organ of smell on pp. 515–19. The numerous changes of wording and rewording here show that Vesalius was not entirely sure of what he had to say, and he is far more concerned with the accuracy of his description. He is particularly interested in how the pathways of sensation are created in even the thickest of bones forming the skull and in what makes any part suitable for admitting odours. He corrects his earlier comment at p. 516, 26 that there is a soft white substance emerging from the brain, writing instead that this is a soft, white process that is drawn out like a nerve. But he goes no further here, promising to say more when he comes to discuss the channels and foramina leading from the nose to the brain in Book VII, but this promise is not carried out. He may have had a special interest in these channels, for his comment that liquid dropped into the ears can sometimes be felt to drip into the cavity of the mouth suggested to at least one modern practitioner that Vesalius had suffered from a perforated eardrum.[113]

The third passage deserves quotation at length. At the bottom of p. 805, Vesalius added two sentences to a discussion of the aqueous humour:

> Because when we are dissecting the eye, the amount of aqueous humour that emerges is small and scarcely comparable with the large amount of vitreous humour, one must conclude that it is largely composed of a sort of spirit and aerial substance that occupies the space between the lens and the pellucid corneal tunic that we generally give to the aqueous humour. Although perhaps someone [might say] that the vitreous humour occupies a larger space in the eye than the rear portion and that the lens along with the vitreous humour is placed off-centre in the forward part of the eye.

Earlier, on p. 798, and again in the introduction to this chapter, Vesalius had been firm in his belief, shared with almost all his contemporaries, that the lens divided the eye into two equal cavities. His successor at Padua, Realdo Colombo, had concluded that the lens was placed a little forward of the centre of the eye, but he did not publish his discovery until 1559.[114] But whether Vesalius' suggestion derives from Colombo or from his own meditation on something he had once found, he does not appear to have backed it up with any new dissection or corrected any of the earlier passages after he had had this thought.[115] Nor does he indicate that some earlier authors had also described the position of the lens as being off-centre.

The absence of any reference to Colombo in the annotations is not necessarily to be explained as a continuation of their rivalry, for the task of reading through the more than 800 folio pages in itself is time-consuming, let alone with the care needed to pick out a missing "i" or replace a comma with a full stop or a semi-colon. At the very least, Vesalius must have spent a great deal of his free time with his book, seeking to improve it and pondering the results of his or possibly others' dissections. This obsessive reading and rereading will have taken months, if not years, for, to judge from the slight differences in the script, Vesalius sometimes went back to change what he had written. It was only death, one might think, that put an end to his obsession to find the right word.

Writing in 2008, before the discovery of these annotations, Jacqueline Vons suggested that Vesalius had taken advantage of the bad weather that had delayed his departure from Venice in 1564 to hand over to a new printer, Francesco de' Franceschi, the copy for a new edition.[116] This was not published until 1568 in a smaller format and with the illustrations redrawn to fit. But Franceschi makes no claim to be publishing a substantially revised edition, but merely to have, at great expense and effort, reduced this "monster" to a smaller size for the benefit of students, especially the poorer ones, and to have had the figures recut.[117] He was telling the truth. He included none of the factual changes recorded in Vesalius' notes, his block-cutter still left out some of the letters that had been obscured in 1555 and, in effect, he reproduced the second edition largely unchanged. He corrected some, but not all, of the mistakes listed in the 1555 Errata, but left some obvious errors unaltered and introduced some of his own.[118] Besides, it is clear from his instructions in his notes that Vesalius still expected Oporinus to be his printer for he alone still had access to the original blocks.

More puzzling than any silence about Colombo is Vesalius' failure to take advantage of the new discoveries by Ingrassia and, particularly, Falloppia.[119] The latter's *Observationes* appeared early in 1561, and Vesalius' reply, later published as his *Examen*, is dated from Madrid in December of that year. In it he discussed many of Falloppia's criticisms and, even allowing for the need to display public courtesy, he appears willing to change his mind on several topics, including the teeth and the structure of the tongue, and to reconsider the role of the fibres in the veins in controlling blood flow, on which he disagreed with Falloppia.[120] By then he had learned, at least by hearsay, of others' discovery of the muscle lifting the eyelid, even if, as he admitted to Falloppia, he still held to a different view.[121] He also knew of Ingrassia's discovery of the third ossicle of the ear, the *stapes* and the stirrup and had confirmed this from his own investigation.[122] Although he comments in his notes on the structure of the ear, and particularly the auditory meatus, he says nothing there about the stirrup bone.

Vesalius' silence about others' discoveries in his notes can be interpreted in two different ways. He may not have known of them when he made his annotations, or he may have deliberately wished to disguise his own failings. The former is more likely, given his willingness to acknowledge some of his mistakes in the *Examen*, and would then imply that the annotations, as we have them, had been completed before 1561 at the latest.

What happened to the volume itself remains a mystery until it reappeared in a German saleroom in 2007. One intriguing possibility is that it may, indeed, have been sent to Basle around 1560 to be printed, but given that copies of the second edition still remained unsold, the arrival of a manuscript of a third edition would not have been welcomed with open arms.[123] The book would have been among the printers' copies and any unsold volumes that Vesalius' family, through their Basle relative Franciscus Maria Stella, attempted to reclaim in 1569 after Oporinus' death.[124] But that does not explain why the volume should have the Walloon (?) version of Vesalius' name stamped on the cover, and a more likely alternative is that these unbound sheets were left behind in his house in Brussels when Vesalius moved to Spain in autumn 1559. Given the fluid political and military situation in the Netherlands at the time, there was every likelihood that Philip II would soon come back there to deal with the Dutch rebels and that Vesalius and his family would not be kept away for too long. Such a scenario would explain both the lack of reference, particularly to Colombo and Falloppia, and the evidence for the survival of the bound volume in a French-speaking environment.

These recently discovered notes add considerably to our understanding of Vesalius and his *Fabrica*. It was never for him a final and definitive statement, but a work in progress, something that he continued to revise obsessively from the very moment of its first publication. Although not every page bears an alteration, few lack a mark that indicates that it has been read at least once by the author, and in some detail. This passion to get things right, whether in the accuracy of the information he provides or in the choice of words, confirms a habit that we have already seen in his attitude towards his two earliest publications, his *Paraphrase* and his *Principles of Anatomy*.[125] His single-mindedness extended not only to the act of cutting and investigating the body but also to the manner in which he proclaimed his message in both words and images.

Three central aspects of Vesalius are here vividly on display. There is Vesalius the anatomist, "we who dissect", not so much presenting new anatomical discoveries, although he continued to carry out at least some anatomical work, particularly involving bones, as ensuring that his readers should not fall into error.[126] Many of the changes, even small ones, are designed to help the potential dissector understand better what Vesalius had done and how he might do likewise.[127] Vesalius thinks as an anatomist. He views the art of dissection as something that requires extraordinary care and precision if it is to be represented on the printed page. The body is something palpable that cannot easily be represented in words.

Secondly, there is Vesalius the humanist, trained from schooldays to write stylish Latin. He corrects mistakes of grammar and syntax (and linguistic errors in his hurriedly written notes are almost non-existent) and changes words and phrases either for clarity or to impress with their style and learning. He is never satisfied entirely, for some of his verbal changes, whether in the 1555 edition or in the notes, may appear trivial or even unnecessary, but he must have felt that he could improve on even his latest formulations. His Latin is not that of Cicero or Celsus but that of his Renaissance co-evals. It is at times ornate, even stately, suited both to expounding complicated technical details and to impressing fellow Latinists, and even an Emperor or two.

Finally, there is Vesalius the artist. His remarkable ability to unite words and pictures and his artistic sensitivity to the layout of the printed page is evident in both printed editions and has often been remarked upon. The notes show that he had both the eye and the firm hand of an artist. A tiny mistake by a cutter is corrected with as much care as a dropped letter, and a thin line is inserted firmly, precisely and clearly.

Defending the *Fabrica*

The period after 1555, as we have already seen from an examination of his annotations, was intellectually challenging for Vesalius. He was largely involved in his court duties and in treating patients and, particularly after his move to Spain, he found it difficult to carry out dissections. This is confirmed by his response to criticisms as expressed in his last published work, his *Examination of the Anatomical Observations of Gabriele Falloppia*, which was written in 1561 and published in Venice three years later. Falloppia, who was responsible for the formal anatomies at Padua somewhat intermittently from 1552, had published his *Observationes anatomicae*, a series of comments on the *Fabrica*, in Venice in 1561 and sent a complimentary copy to Vesalius.[128] It was different in both tone and content from previously published critiques of the *Fabrica*, except for Massa's, being the response of an experienced dissector who had had long experience of following Vesalius's example (Fig. 4.8).

These *Anatomical Observations* take the form of extended comments on the *Fabrica* based on some twenty years of dissecting and practising surgery, with Vesalius replacing Hippocrates or Galen as the basis of the enquiry. It is a courteous presentation of their differences, praising both the author and his book as "divine" and declaring that this extremely important work would last forever.[129] But it was not perfect, and Falloppia hoped to steer a middle course between Galen, whose expertise

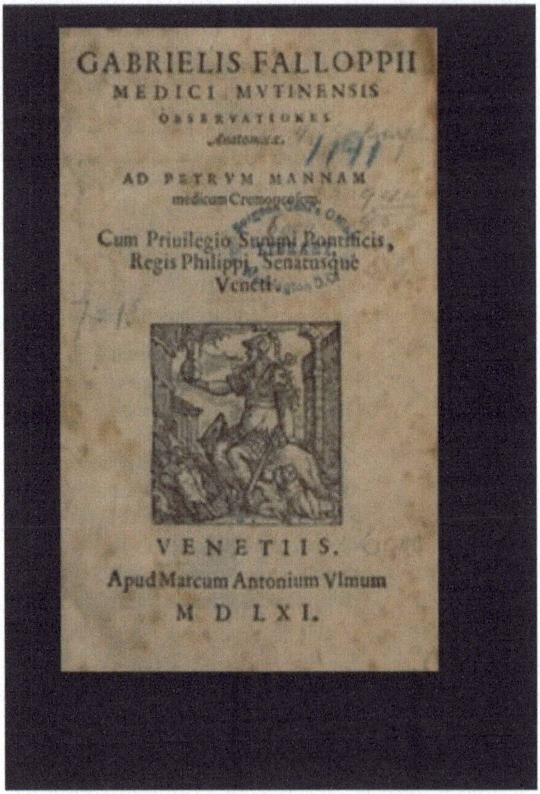

Fig. 4.8 Gabriele Falloppia, *Observationes anatomicae*, Venice: Marcantonio Olmo and Gratioso Perchachino, 1561, title page. (*Source* Wellcome Collection)

with the knife he acknowledged, and Vesalius. He also noted what had been a major change since 1543, the increasing number of anatomists with experience of dissection, the great majority of whom were now following Vesalius.[130] As the Venetian doctor Girolamo Donzellini puts it in a letter to a friend at the end of the 1550s, a few opponents might still remain in Italy and elsewhere, but, like the monsters slain by Hercules, they were no longer formidable.[131] Things that had been unclear in 1543, and hence deeply divisive, were now clarified because of the increased amount of evidence now becoming available and being communicated to other anatomists verbally or by letter. The third ossicle in the ear was discovered by Ingrassia in Sicily and confirmed by Falloppia's own dissections, but the news of this find had apparently reached Vesalius before he read the *Observations*.[132] The art of illustration had also improved, and Falloppia planned a much larger volume with many illustrations of both animals and humans that would include material provided by Galen, Vesalius and others (although in the event his unexpected death cut that project short).[133]

Falloppia continually stresses how his own considerable experience had led him to a better understanding of anomalies and variations in the human body.[134] He had learned from his many dissections of the corpses of young children that the developmental processes in the very young proceeded at different speeds, and some sutures in the skull or jaw did not coalesce for months, if not years, after those in others.[135] He had had also access to innumerable "huge heaps of bones" in Ferrara, Florence and elsewhere on his travels so that he had a better idea than Vesalius of the range of possibilities for the human skull.[136] Likewise, in his view, a sacrum composed of six pieces was more common than one of five, the number on which Vesalius and Galen had both agreed.[137] Some of the divergences between Vesalius and his opponents might simply be the result of misunderstanding Greek terms—or, in fact, because the ancient Greeks themselves were divided about terminology.[138] At times modern anatomists disagreed among themselves: Vesalius believed there were nine muscles moving the tongue, Valverde ten, but despite carrying out several dissections to decide between them, Falloppia failed to find more than eight, and believed that both had wrongly classified other phenomena.[139] As for the valves in the veins that Amatus Lusitanus had described on the authority of Canani, Falloppia was keen to associate Vesalius with his own refusal to believe in them, for they were not mentioned in either edition of the *Fabrica*.[140] Throughout, his reservations are expressed courteously and with constant reference to his own experiences and those of others as dissectors or informants, as well as to the difficulties involved in seeing and classifying some of the tiny structures within the body. It is a far cry from the shrill attacks of earlier Galenists.

He was more critical in his lectures to his students, referring often to Vesalian errors and to the many times in which he had been deceived in basing his human anatomy on that of animals (including monkeys as well as dogs). He had been less diligent than he should have been in describing the abdominal muscles and in supposing that the *vena humeraria* branched off from the *vena axillaris* (which was true in monkeys) rather than both branching next to each other from the *vena cava*. He also emphasised his own greater experience with the corpses of both criminals and patients in hospitals. But these comments were muted in the published text, although the sharp-eyed Vesalius might well have noted that among the kind words Falloppia used to refer to him, it was Berengario, not himself, who was credited with being the initial restorer of the art of anatomy.[141]

Vesalius responded quickly to Falloppia, but the Venetian ambassador to Madrid, Paolo Tiepolo, to whom he had entrusted his reply, was delayed for several months in Catalonia by the religious struggles in France and by the lack of a suitable boat home, only to find that by the time he reached Venice Falloppia was already dead. Fortunately, he held on to the manuscript, and after Vesalius, during his brief stay in Venice in 1564, had explained the reasons for its non-appearance to Gadaldino and other well-wishers meeting him at the bookshop of the printer Francesco de' Franceschi, arrangements were quickly made for its publication[142] (Fig. 4.9). De' Franceschi's preface bears the date of 24 May, roughly two months after Vesalius left Venice on an official voyage to the Holy Land to bring the annual subsidy from the King of Spain to the churches there. The book takes the form of a private letter, but one that may have been intended to be circulated among the cognoscenti, although not necessarily published immediately.[143] Indeed, Vesalius had even raised the possibility with Falloppia that his comments might be published alongside his *Observationes* in a future edition. Vesalius acknowledged his responsibility for some of the mistakes that had been identified, admitting that since his happy days in Padua, his time had been taken up with surgical aspects of medicine, attendance in constant wars and perpetual travel. He could see no foreseeable chance of performing dissection in Spain, where he could scarcely obtain a skull, although he hoped that some further opportunity might be granted him.[144] He was not ashamed to have laid the foundations of the many new developments in anatomy and hoped that he could be kept abreast of Falloppia's latest discoveries.[145] But that was for the future, and his reply to Falloppia merely presents his immediate reactions to his news and throws light on some of the circumstances in which his masterpiece was written.

In the *Examen* Vesalius is far less combative than before. He shows considerable respect for the younger man, at least at the beginning and end of his reply, and any criticism of his work as an anatomist is at best oblique. His reference to his own autopsy of a man who had previously consulted Falloppia and others for some painful nodules around the penis occurs during a passage where Vesalius thought that he had been misunderstood by Falloppia.[146] He makes a more obvious correction when

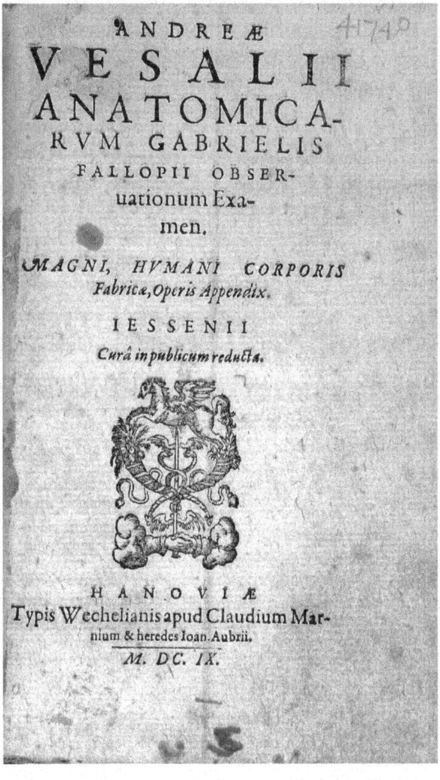
ANDREÆ
VESALII
ANATOMICA-
RVM GABRIELIS
FALLOPII OBSER-
uationum Exa-
men.
MAGNI, HVMANI CORPORIS
Fabricæ, Operis Appendix.
IESSENII
Curâ in publicum reducta.
HANOVIÆ
Typis Wechelianis apud Claudium Mar-
nium & heredes Ioan. Aubrii.
M. DC. IX.

Fig. 4.9 Andreas Vesalius, *Anatomicarum Gabrielis Falloppii observationum examen*, Hanover: Wechel Press, 1609, title page. This edition (originally 1564) was curated by the Wittenberg anatomist Johann Jessen (1566–1621) as an appendix to the *Fabrica*. (*Source* Wellcome Collection)

discussing the role of "our mutual friend" Canani in identifying a valve in the azygos vein, declaring that Canani could not have described it to Amatus as a mocking joke against him, for he had also been told about it by its discoverer.[147] He is more robust in rejecting Falloppia's claims to have discovered the clitoris, which for him was a sport of nature, something only to be found in hermaphrodites, a pathological anomaly.[148] But old enemies do not escape so lightly. Valverde is accused of never dissecting himself and being ignorant of medicine as well as the logical and linguistic skills that lie at its base. Falloppia should beware of placing any trust in his authority.[149] Besides, neither he nor his master Colombo, who had learned almost everything from Vesalius, was sufficiently well versed in Galen to appreciate what he had done, let alone to check in the manuscripts to identify a later unwarranted change to what Vesalius had emended in 1541. This was what he would have done immediately, had he had the relevant material to hand in Spain.[150] After all, when he managed to find in Spain a copy of Galen's *On the Dissection of Muscles,* he used it at once to see what Falloppia had said about the *teres* muscle.[151]

His response to Falloppia is almost universally positive, at least on the surface. He judged his description of the teeth to be "extremely satisfactory", and he explained that he had already rectified his earlier ignorance of the stirrup bone in the ear by confirmed Ingrassia's discovery for himself.[152] Other items, such as the *palmaris brevis* muscle and the valves in the veins, he had already observed but classified them differently.[153] His belief in the six-boned sacrum advocated by Galen had been strengthened by his experience of such a one belonging to a middle-aged man that he had observed while in Padua, but he was now aware that this was something to be found in dogs and apes.[154] In his early work on the sutures of the skulls he had been led astray by what Hippocrates had said, but he had since learned much more about the different types of skulls from examining specimens in the cemeteries of Styria and Carinthia in May 1552.[155] He admitted his misunderstanding of the meaning of the word *ginglymus*, although back home in Brussels they used hinges of a different kind, and, in a humorous aside, he noted that Spanish doors were almost always hung from wood rather than iron.[156]

Particularly significant is his acceptance of the existence of a muscle lifting up the eyelid, which, as we have seen, he was still denying in his notes to his second edition.[157] Galen, and following him Oribasius, had suggested that there was a muscle going from the frontal bone to the eyelid, but all that he could find, after several attempts, was some glandular flesh and a small fragment of muscle that he believed was part of the muscle that moved the eye itself, not the lid. He thus considered that what he had found in the eyelid itself was some sort of cartilage. "Provoked by Roman diligence" (an oblique reference to the hated Colombo and possibly Eustachi (d. 1574)), he carried out at least one more dissection of the eyelid, finding something round and fleshy that might raise the eyelid. It had an unusual make-up, something between the flesh of muscle, a glandular substance and the thick fluid around the joints. This he termed a gland rather than a muscle but was now prepared to accept Falloppia's correction with reservations. His willingness to respond to new suggestions by attempting to confirm them by a dissection of his own is noteworthy, even in the limited facilities at his disposal in Spain, and he may have drawn on his experiences of autopsies and premature births to offer a few new observations of his own.[158]

This is the last clear evidence we have of Vesalius the anatomist. But there were already rumours circulating about his wish to return to academic life at Padua even before he came to Venice on his way to the Holy Land. The teaching of anatomy and surgery in Padua had suffered two devastating blows with the deaths of Falloppia and Landi in quick succession in 1562, the former from plague, the latter murdered in broad daylight in mysterious circumstances.[159] There was no obvious successor to Falloppia. There was an anatomist called Francesco Lendinara, but he was a nonentity, and the junior medical professor Alvise Bellocati was put in charge of the annual anatomy.[160] By mid-March 1563 no formal anatomical and surgical teaching had taken place, and there were even fears for how the authorities might respond now that the university was a mass of mediocrity, a frozen academic waste. There were rumours that Vesalius was going to return, but, as one student remarked, this was just a rumour.[161]

The failure to fill the anatomy post adequately for a long time may have only added to speculation about what Vesalius was planning when he spent time in Venice on his way to the Holy Land and met with several leading physicians.[162] As he wrote to Falloppia, he still had not given up hope in late 1561 of being able to return to dissecting, even if at court he was subject to severe constraints, and a resumption of an academic career in as pleasant an environment as Padua would have had its attractions. There was time enough for discreet discussions with the Venetian authorities who oversaw the university, and Pietro Bizzarri, who spent much of 1565–6 in Venice and who was acquainted with Benedetto Varchi, a friend and correspondent of Vesalius, reported that only a few days after he had set out for the Holy Land the Venetian Senate invited him back to Padua at a substantial salary.[163] But no trace of this decision has yet been found in the archives or in any of the many student letters that were sent from Padua back to homes in Germany. Philip II certainly expected him back in Spain, and an unauthorised departure from his post at the court of the most powerful ruler of the day was not without its dangers.[164] This story may be as unlikely as the claim that he had fallen under the suspicion of the Inquisition and had made his voyage as a penance for having accidentally performed a human vivisection.[165] This is highly unlikely, a reflection of common fears about the increasing persecution by the Inquisition in Spain and elsewhere and about dissection at the hands of ruthless anatomists.[166].

Even if Bizzarri was right, Vesalius never had the opportunity to take up his old chair. His return journey from the Holy Land was a catastrophe. The boat on which he set sail was allegedly driven before a storm for forty days without landfall, provisions ran short and Vesalius fell ill. The ship finally made land on the island of Zante (now Zakynthos) off the West coast of Greece, and Vesalius came ashore, only to die almost immediately. His end is as mysterious as his beginning. Two alleged eyewitnesses offer different stories. One, a Venetian, has him dying of plague in a remote inn, and being buried by a compassionate goldsmith; the other, more plausibly, puts his death closer to the main town of Zante, where a German erected a monument in his memory. Later antiquarians recorded an inaccurate epitaph from the Catholic church in the main citadel of Zante, but this has long disappeared, if indeed it ever existed.[167]

Vesalius' wife, having received a pension from Philip II in view of the long and devoted services of her late husband, soon married again, to Henrich Vandermeeren, the Seigneur of Saventhem, and enjoyed a further thirty years of marriage. His daughter, Anne, married and lived comfortably in their old house until 1587, when it passed to their neighbour, Count Mansfeld. The last descendant of her five children died in the mid-nineteenth century at Courtrai, Belgium, having also been in service with an Emperor, Napoleon.

The Later History of the *Fabrica*: The Wooden Blocks

The history of a famous book can take various forms, from a listing of its editions or its surviving copies to an examination of readers' reactions to it and an analysis of its wider impact whether bibliographical, scientific or literary. For the *Fabrica*, we can not only trace the author's own struggles to improve what he had originally written, as we have seen, in both its published form and in his marginal notes, but also follow the history of the wooden blocks for the illustrations from their creation in Venice in 1542 until their destruction in Munich during the Second World War.

The original drawings, by a variety of hands, were transferred to the blocks and then cut probably in the Venetian workshop of Francesco Marcolini.[168] Carefully packed and transported to Basle, they may have reached there in September 1542, but the complicated task of setting them took time, enough for Vesalius to have second thoughts about two of the historiated initials.[169] Most of them were reused again for the 1555 edition and were regarded as sufficiently valuable, especially if a third edition was ever to be published, to be specifically mentioned in the 1568 lawsuit between the heirs of Vesalius and those of Oporinus.[170] Twelve years or so later, when the Basle anatomist Felix Platter, the son of Oporinus' old partner, was seeking illustrations for his own anatomy book, he had the chance to buy them from the heirs, but he was deterred by their sheer size, which was inconvenient for students. No one, in his view, had ever made better images of the body, but it was easier to use smaller copper-plate engravings of them as others had done.[171] It is unclear whether heirs Platter was referring to the heirs of Vesalius, for the lead plaintiff in the case, Vesalius' cousin Franciscus Maria Stella, was living in Basle as a citizen, or those of Oporinus. But there are no grounds for assuming that Platter's publisher ever owned the blocks, for he could easily have engaged an artist to copy from the book, as others did, including Valverde and de' Franceschi.[172]

All traces of them then disappeared for over a century until some of the images came "by remarkable chance" into the possession of an Augsburg printer and publisher, Andreas Maschenbauer (1660–1727). In 1706 he selected nineteen of them, drawn from both the *Fabrica* and the *Epitome* and confined to the external appearance of the body and its parts, to illustrate a small guide for artists and sculptors (Fig. 3.7). The title page bears the following title: *Andreae Vesalii Bruxellensis, Deß Ersten,*

Besten Anatomici, Zergliederung Deß Menschlichen Körpers. Auf Mahlerey und Bildhauer-Kunst gericht. (*The Organisation of the Human Body, by Andreas Vesalius the First and Best anatomist, a Contribution to Painting and Sculpture.*) At the bottom of the page is the (erroneous) information that these images were drawn by Titian. In following the trend set by Tortebat and Domenico Bonaveri (Fig. 4.10) a generation earlier to use Vesalian images as the standard for intending artists to copy, Maschenbauer now emphasised that he had reproduced his illustrations by Titian "just as they had appeared in the hands of their original owners".[173] Here and in a second edition seventeen years later, he emphasised the fame of Titian and his status as an imperial artist, adding further to the significance of Vesalius.[174]

Around 1774, the Bavarian physician Johann Anton von Wolter (1709–87) rediscovered and purchased the blocks, probably from Maschenbauer's heirs, and brought them to Ingolstadt, where he was Dean of the medical faculty of the university. His plan was to print from them and, with German captions, to illustrate a small introductory manual of anatomy for surgeons, largely taken from a more up to date manual by Jacobus Benignus Winslow (1669–1760).[175] He was helped in this by his younger colleague Heinrich Palmaz Leveling (1742–98), a professor of anatomy, physiology, and surgery at the university. An attempt to publish the book by subscription failed, and the Leipzig publisher Siegfried Leopold Crucius (1738–1824), although initially interested, baulked at the task when confronted in 1777 with the number and size of these old wood blocks. It was not until 1781 that Leveling managed to collect enough subscribers to begin printing parts of the volume with the Ingolstadt University printer, and not until 1783 that the whole treatise saw the light of day.[176] By then several of the blocks had been lost or replaced because of cracks and other faults.

Following Wolter's death, the blocks were deposited in the University Library at Ingolstadt, and their fate from then on was linked to the history of that university and its collections. It was twice moved, first to Landshut, somewhat closer to the Bavarian capital of Munich, to which it was finally transferred to more majestic buildings in 1826. Such moves of both buildings and personnel are always likely to lead to books and materials being lost or, more often, neglected, particularly when stored in boxes of uncertain origin. Such was the fate of the blocks, for when in 1885 Moritz Roth (1839–1914), the Basle professor of pathology and the leading expert on Vesalius in his day, wrote to Munich University, having failed to find any

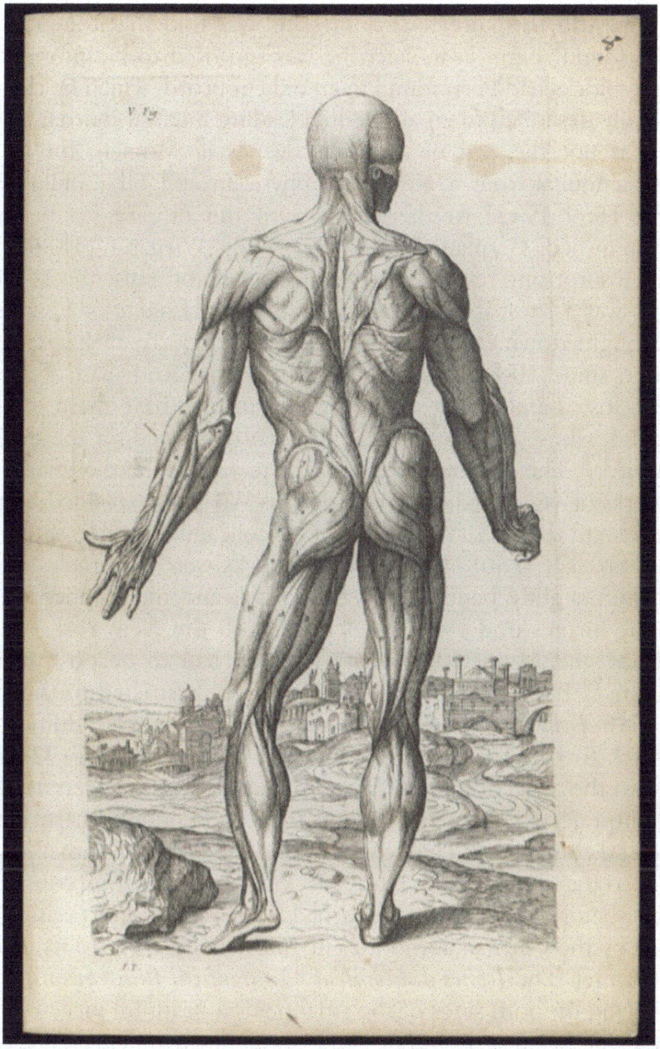

Fig. 4.10 François Tortebat's use of Vesalian images in his *Abregé d'anatomie accommodé aux arts de peinture et de sculpture*, Paris: Tortebat, 1667, was widely imitated in later handbooks for artists, partly because they were claimed to have been drawn by Titian. (*Source* Wellcome Collection)

trace of them in their previous homes, he was told that they could no longer be found. Eight years later, he was surprised to be informed that 135 of the blocks had been found in an old cupboard, which he examined and carefully described in an article in a leading medical journal.[177]

This was not the end of the rediscoveries in Munich, for in 1932 a hopeful enquiry from a New York physician and bibliophile Samuel Lambert (1859–1942) to the Director of the Library via his friend, the distinguished German typographer Willy Wiegand (1886–1961), produced a surprising result. The Director, Dr Adolf Hilsenbeck, recalled that there was a chest in the attic marked with the name of Vesalius and had it brought down to be examined by Wiegand. To their amazement they found some 30 blocks, including many that Roth had declared to have been lost. Lambert set about raising funds to have them published by Wiegand, whose private Bremer Press was renowned for the quality of its printing.[178] The original intention was to reproduce the images from all the works of Vesalius, a huge task that, as Wiegand explained, required special care and the remarkable skills of his colleague as cutter and printer Joseph Lehnacker (1865–1965) for the blocks were extremely fragile as well as often slightly bent. It also took time, and the planned publication of the images under the auspices of both the New York Academy of Medicine and Munich University in 1934 had to be postponed for over a year.[179] Its wider ambition had also to be restricted to the images of the 1543 *Fabrica* and *Epitome*, with one remarkable addition. The wood block for the title page of the 1555 edition had by 1924 found its way into the collection of an Antwerp banker, Laurent Fierens (1875–1961), who rejected an offer to buy it from the London pharmaceutical magnate and collector Sir Henry Wellcome (1853–1936) so that he could present it to the library of the University of Louvain. It took considerable diplomacy before the library, which had been rebuilt after being totally destroyed by the German authorities in 1914, could agree to its loan to a Munich printer. The *Icones anatomicae, Anatomical Illustrations,* greeted with rejoicing on both sides of the Atlantic, is a beautiful piece of typography, and the *Vesalian Essays* of 1952, some of them delivered earlier in New York as speeches marking its appearance, are still fundamental in any discussion of the Vesalian images. Its publication came just in time. The 1555 block was lost in May 1940 when German artillery set the Louvain library on fire once more. The fate of the Munich blocks is perhaps even more tragic, for although many of the treasures of the University were sent away to a distant store, its Director, obeying a government order to keep the University working as normally as possible, kept many back to put on exhibition and for scholarly use. They were stored in a "bomb-proof" basement which survived one attack by American bombers in May 1944, only to fall victim to a second few days later. All traces of the blocks vanished in the ashes of their last home.[180]

The story of the blocks is not only significant for the history of book production, as the essays of Wiegand and Goree, in particular, have shown, but it also demonstrates the importance of the images and the varied uses to which they were put over the centuries. In 1590 the Basle anatomist Caspar Bauhin (1560–1624) proclaimed as a selling point on the title page of his anatomical treatise that it was based on his lectures and on Vesalius' illustrations.[181] Since then, they have circulated as images on their own or accompany unauthorised summaries of the *Fabrica* as well as much later treatises on anatomy, like that of Tortebat (Fig. 4.10); some have even been set in marquetry panels and into baroque marble tomb slabs in a Maltese cathedral.[182] As early as 1560, the Lyons printer Clément Baudin was including artists and architects among the projected readership for his selection of images from Vesalius. (One of the two surviving copies was owned by an apothecary, the other by a French aristocrat.)[183]

Long after many details of Vesalian anatomy had been superseded, the images retained their power as objects of beauty (helped by the belief in their creation by Titian).[184] Even today, far more of those interested in the *Fabrica* know it through them, and especially through Saunders and O'Malley's 1950 *The Illustrations from the Works of Andreas Vesalius of Brussels* and its subsequent inexpensive reprints in both hard and paper back, more than through the *Fabrica* itself.[185]

It was not always thus. The Lyon 1552 edition, published as soon as possible after the ending of the French royal copyright, contains only 4 illustrations, and the emphasis in its two duodecimo volumes is on the word. What Vesalius had written was now made available at a price that more students could afford, and in a new, more convenient size for carrying around.[186] Given that versions of at least some of the plates were already in circulation cheaply, either in the form of fugitive sheets or in versions of the *Epitome* deplored by Vesalius, and were soon to be followed by the celebrated Plantin printings of Valverde and Jacques Grevin, the decision not to reprint them is not as foolish as Cushing believed.[187] In 1551 Vesalius' erstwhile friend and supporter, Leonhard Fuchs, found another way of delivering the message of the new anatomy to students by producing his own *Epitome* of the *Fabrica*. In two octavo volumes of some 490 pages, it steered a middle path between the brevity of the Vesalian *Epitome* and the massive *Fabrica* (and, as the title page proclaimed, between Vesalius and Galen). Like the 1552 edition it eschewed illustration.[188]

The Later History of the *Fabrica*: From Textbook to Icon

The sheer size of the *Fabrica* was one of the reasons put forward by the editor of the fourth printing, Francesco de' Francheschi, Franceschi, the printer of the *Examen*, in the preface to his edition of 1568 (Fig. 4.11). In two volumes, in smaller folio format, it reproduced all the images of the 1555 version, along with a new title page, but, of necessity, all had to be recut and reset. The images were remade by an artist from Pomerania, N. Germany, Johann Crieger, whose name appears on the title page alongside that of de' Franceschi, and follow very closely those of 1555.[189] But although the printer had known Vesalius, there is no evidence that this edition had been authorised by Vesalius or that he knew of the revision Vesalius had made in his own copy. It was a commercial speculation and successful enough for two of the printer's sons to bring out yet another edition in 1604, simply entitled *Andreae Vesalii Anatomia*[190] (Fig. 4.12). They reset the whole text, but so carefully that everything matched its predecessor line for line. They also added in a separate appendix volume three Ramist tables, tree diagrams setting out the ancients' ideas on the parts of the body, taken mainly from Rufus of Ephesus and a fourth, on the womb and female genitalia, taken from Soranus.[191] They had been devised by the anatomist Fabio Paolino as an aid to learning for his students and, so it was hoped, would provide a simple way of remembering what was in the *Fabrica*.[192] Whether Vesalius would have approved of this addition is uncertain, but his critical eye would have quickly noticed that, as Monique Kornell reminded me, some of the torsi in the niches on the title page are taken from his pet hate Valverde's slightly different modifications of those in Book 5 of the *Fabrica*.

The influence of the *Fabrica* can also be seen in the most important of sixteenth-century treatises on veterinary medicine, Carlo Ruini's *Anatomia del cavallo, infermità, e suoi rimedii, The Anatomy of the Horse, its Diseases and Treatments*, first printed in Bologna shortly after his death

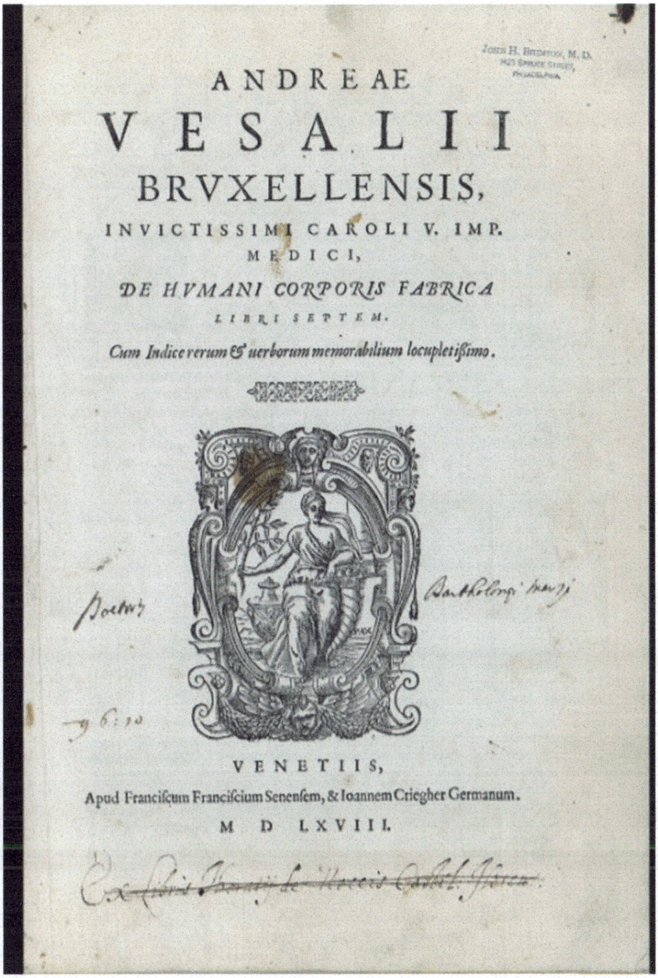

Fig. 4.11 The title page of the Venice edition by Francesco de' Franceschi, 1568, who reduced the size of this 'monster' and had the images redrawn. (*Source* Wellcome Collection)

Fig. 4.12 Vesalius, *Anatomia*, Venice: Giovanni Antonio and Giacomo de' Franceschi, 1604, title page. This is the last edition specifically prepared as a textbook of anatomy. (*Source* Wellcome Collection)

in 1598 (Fig. 4.13).[193] A substantial treatise in two books, its first book, dealing with the anatomy of the horse, provided for veterinarians, as well as a wealthy readership, a detailed account that was modelled in many ways on the *Fabrica*. Not only was it illustrated with large-scale images of individual parts of the anatomy of the horse, several reminiscent of those of Vesalius, but the four full-page ones showing the whole horse, like those of the musclemen in the *Fabrica*, set the horse against a similarly idealised background. Like the images in the *Fabrica*, they were endlessly copied in later manuals of veterinary anatomy.

The 1604 edition of the *Fabrica* marks a turning point in its history, for as Cushing's list of Vesaliana shows, it effectively marks the end of the authority of the Vesalian text as new discoveries in anatomy made it ever more outdated.[194] The full text of the *Fabrica* was only reprinted once more before the production of modern facsimiles of the original in the twentieth century, in 1725, as part of a lavish edition of the complete works of Vesalius[195] (Fig. 4.14). Beautifully printed and with its images newly engraved, it was edited by one of the greatest names in Dutch medicine, the physician Hermann Boerhaave (1668–1738) assisted by the up-and-coming anatomist Bernhard Siegfried Albinus (1697–1770) at a time when the reputation of Leiden and its medical school was at its height. It appears to have sold well and many copies remain today, but it is far from clear how far it was read and used, and it may have been acquired more as a mark of their owner's status and respect for the most famous of all anatomists than to be consulted regularly.[196] As David Soulier remarks, this is a work of hagiography, for the editors in their preface say next to nothing about the value of their edition for students or for libraries eager to acquire everything that a great writer had written, both common themes to justify publications of an author's complete works.[197] Instead, they provide their readers with a long and detailed biography of this man of genius that still resonates today with many medical professionals. They stress his precocity, his willingness to challenge authority, his quarrel with Sylvius, his unhappy life at court under the pressure of the Inquisition and his sad death on Zakynthos. Recent scholarship has modified, if not rejected, many of their conclusions, but their overall message of a medical hero remains influential.

The twentieth century saw two new developments in the life of the *Fabrica*. Technical changes in methods of printing, as well as a substantially larger market of potential buyers among both medics and connoisseurs, led to the publication of full-size facsimiles in 1964, 1970

Fig. 4.13 Carlo Ruini, *Anatomia del cavallo, infermità, e suoi rimedii*, Venice: Fioravante Prati, 1618 (originally 1598). The new plates in the second edition make its dependence on Vesalius even clearer. (*Source* Wellcome Collection)

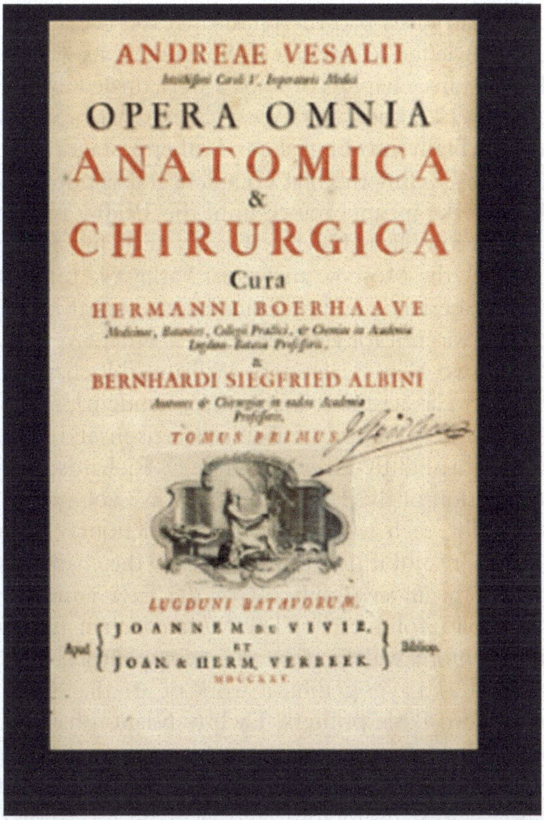
ANDREAE VESALII

OPERA OMNIA
ANATOMICA
&
CHIRURGICA
Cura
HERMANNI BOERHAAVE

&
BERNHARDI SIEGFRIED ALBINI

TOMUS PRIMUS.

LUGDUNI BATAVORUM,
Apud JOANNEM DU VIVIE, ET JOAN. & HERM. VERBEEK. Bibliop.
MDCCXXV.

Fig. 4.14 The title page of first volume of the collected works of Vesalius, published at Leiden by J. de Vivie and J. and H. Verbeek in 1725, and edited by the celebrated Leiden physician Herman Boerhaave and the younger anatomist Bernhard Sigismund Albinus. (*Source* Wellcome Collection)

and 1975, as well as smaller selections, too numerous to mention, of some or all of the plates.[198] Before 1950 it was also normal for anyone with a medical degree to have some expertise in Latin, even if not always at the level to cope with so long or complex a book, a factor which always inhibited its use from 1543. But with the growing demise of Latin, particularly in anglophone countries, came a need for translation into an accessible vernacular. The sheer size of the *Fabrica*, however, presented a serious barrier, and few anatomists had the linguistic expertise and few Latinists the anatomical knowledge to attempt it. Some sections were published in translation. The first five chapters of book I were turned into French by L. E. Verschaffelt in 1924, almost thirty years later, Charles Singer translated the sections on the brain into English, and other passages were also made available in English.[199] But the first complete translation into a modern language did not make its appearance until the 1950s, as part of a series of Classics of Science published by the Russian Academy of Sciences.[200] It was a joint effort by the Moscow anatomist Vasiliy N. Ternovsky and the classicist Sergei P. Shestakov and included an afterword by the physiologist Ivan Pavlov. It was not for another fifty years that a version in English was produced, by two New Zealanders, William F. Richardson and John B. Carman (another classicist and anatomist in tandem), followed quickly by the similar American pairing of Daniel Garrison and Malcolm Hast. Each pair followed a slightly different strategy. Richardson and Carman published their version of the 1543 *Fabrica* in five volumes that are relatively easy to handle, each with its own introductions.[201] Garrison and Hast were much more ambitious, translating also the changes in the 1555 edition as well as the most significant of the later annotations in two large and heavy folio volumes.[202] Their *Introductio*n is more detailed and ranges much more widely than their competitors, and the whole book was designed as a typographical work of art that would show off the quality of modern Basle printers. Each translation has its merits, and both have also enhanced the legibility of the original images with the aid of modern technology. Together they make available the whole of the book in English, at least in libraries, and facilitate the study of Vesalius' arguments by the non-Latinate. A larger French group is in the process of producing a French translation of the 1543 *Fabrica*, and much else, online on the website of the Paris Bibliothèque Interuniversitaire de la Santé, which makes their version even more accessible to scholars at the click of a mouse.[203] Having the possibilities of having texts, translations and transcriptions of most of the editions available together on the screen opens up new ways of consulting the *Fabrica* that were impossible a little more than a decade ago.

Owners and Users

But who owned the *Fabrica* and what use did they make of it? Until recently historians have had to rely on the information provided by Cushing in the revised *Bio-bibliography* and on the occasional detailed description of individual copies by bibliographers such as Elly Cockx-Indestege.[204] But the remarkable *Census* of Daniel Margócsy and his colleagues, supplemented by the thesis of David Soulier and his smaller census of copies of the *Examen*, has opened up new lines of research into the reception history of the *Fabrica*.[205] While it is generally agreed that by the end of the sixteenth century Vesalius was firmly established as the creator of modern anatomy, we now have a clearer vision of the ways in which the book was owned and interpreted over almost five centuries.

The Census of copies of the 1543 and 1555 editions lists 309 of the former and 438 of the latter, including those in private hands and those lost or destroyed.[206] This compares with some 95 examples of the Latin *Epitome* and 30 of the German version, perhaps because they were regarded as less impressive or less valuable than the complete work, but more likely because also they were used more heavily.[207] Some copies were specially prepared for presentation to possible patrons. The copy of the *Epitome*, now in the Cambridge University Library, presented almost certainly to Philip II, and that of the 1543 *Fabrica*, once part of the Haskell Norman Collection, that was given to Charles V, are magnificently coloured and with splendid bindings, presumably commissioned by Vesalius himself and prepared by Oporinus.[208] Ironically, the latter was valued more as an *objet d'art* for use in diplomacy, for the Emperor passed it on within a few years as a gift to Jacques Mesnage (1509–1556) on leaving his post as the French Ambassador at his court in 1549. Other copies were given to friends by the author. The son of Renatus Hener wrote into the family copy of the 1555 edition that it had been given by Vesalius to his father in return for his defence of Vesalius against Sylvius, below, p. 219, note 36.[209] Others were donated to colleagues with wide contacts, such as the Augsburg physician Achilles Pirmin Gasser (1505–77) or the friend of Oporinus, the Swiss humanist and physician, Joachim Vadianus (1484–1551).[210] Oporinus had sent him a complimentary copy, probably of the 1543 *Epitome*, as soon as it was printed, but the carrier's horse was said to have slipped into a stream on the way to St. Gallen and the book was lost. Somewhat suspicious of the carrier's story, Oporinus nevertheless had to send a replacement, the copy of the *Epitome* which now is one of the treasures of the St. Gallen Stadtbibliothek. But why Vesalius should have given a copy in 1548 to Stephanus Lagus, whom he knew briefly at Esslingen, is unclear. Another volume that has been thought to have been given by Vesalius, once owned by Jan Viringus (1539–1598), the later Dutch translator of the *Epitome*, and heavily annotated by him, is more likely to have come into Viringus' possession later without Vesalius' intervention.[211]

Surviving copies of both editions show that for many decades most were owned by medical men and academics of high status. Physicians outnumber surgeons, and royal surgeons like Nicholas Rasse Deneux (d. 1581) ordinary surgeons, reflecting their difference in wealth or status, and some institutions, such as Oxford Colleges, were early purchasers.[212] Other copies were owned by bibliophiles, such as Johann Friedrich, the Elector of Saxony and Anna Jagielonka, Queen of Poland, as well as artists such as Alessandro Allori (1535–1607), who had a particular interest in anatomy.[213] There will also have been many early owners whose copies have either been lost or cannot as yet be identified, but who can be presumed to have received a complimentary copy, such as Cardinal Granvelle, one of Vesalius' friends and patrons at court, or whose ownership or use is mentioned in a contemporary letter or notebook. So, for example, soon after 1543, it was being studied by Juan Páez de Castro (c.1512–70), the agent of the Spanish ambassador to Venice and celebrated bibliophile, Diego Hurtado de Mendoza (1503–75), although it is not clear whether he or, more likely, his master was the owner.[214]

Although details of modern ownership are a weak guide to the early distribution of the first two editions of the *Fabrica*, it is clear that they sold widely and across the confessional divide. Alongside the reformed ministers of Zurich, Heinrich Bullinger (1504–75) and Rudolf Gwalther (1519–86), can be set the Jesuit bishop of Sassari, Andres Bacallar (1515–1612).[215] There is a substantial concentration of owners in Germany, reflecting not only the large numbers of German students who crossed the Alps to study and the wealth of its towns and small courts, but also the influence of Melanchthon and Wittenberg. Melanchthon not only used the *Fabrica* heavily in the second edition of his tract on the soul, a major influence on Lutheran thought, but wrote a laudatory poem about it in his own copy.[216] Several of his colleagues and pupils at Wittenberg owned their own *Fabrica*, sometimes using it as the basis for lectures within and without the medical faculty (above, p. 162).[217] So familiar was it that the head of Vesalius could be used on the male figure of one of the anatomical fugitive sheets that were bought as an accompaniment to lectures on Melanchthon's treatise on the soul (Fig. 4.15). Elsewhere, one owner, who went carefully through his copy of the 1555 edition, wrote out in the front the opening passage from a 1578 lecture by the Rostock professor and theologian, David Chytraeus (1531–1600). In it, Chytraeus, whose brother also owned a copy of this edition, apologised for the forthcoming replacement of his next week's lectures by a public dissection and encouraged his students to attend diligently to all that they would find there. Far more learned and more pleasing than anything he could offer, the "anatomical lecture in theology" would reveal the plan and supreme artifice of the Creator in devising the human body.[218] Younger anatomists in some Spanish universities seem also to have welcomed Vesalius' ideas, even if elsewhere there was a great reluctance, and it was a Spanish physician who transported to Mexico what is almost certainly the copy that has remained longest in the Western hemisphere.[219]

Its owner kept up to date in his anatomical studies, at least for a while, adding references in his margins to Colombo, Falloppia and especially Valverde. Like many other owners, he was particularly interested in comparing what Vesalius had said with Galen, but without taking any obvious position on their disagreements. Open hostility to Vesalius is rarely found, but the unknown reader who had attended a dissection in Padua in 1574 was typical in being much more at home with his classics, Galen, Hippocrates and Aristotle, than with the day-to-day practice of surgery or with dissection. Like most annotators, he was interested in what the *Fabrica* could contribute to wider medical discussions about the workings of the body.[220] Most readers underlined passages rather than commenting verbally on what they found, and rarely went beyond the occasional correction. The owner of the Johns Hopkins copy of the 1543 edition is unusual in that he sedulously went through the preface writing in all the changes that Vesalius had made in his revision, but the task was beyond him and his interventions gradually diminish and finish entirely before reaching the end of Book I.[221] Several readers used their copy to record their own opinions and experiences, whether dissecting themselves or attending the dissections of others. The owner of the Poznan copy twice records his presence at dissections by Vesalius in 1543, including seeing his celebrated dissection of the monk's mistress.[222]

Annotations can also take us into the lecture room. Two students recorded details of an autopsy and dissections taking place at Leipzig in

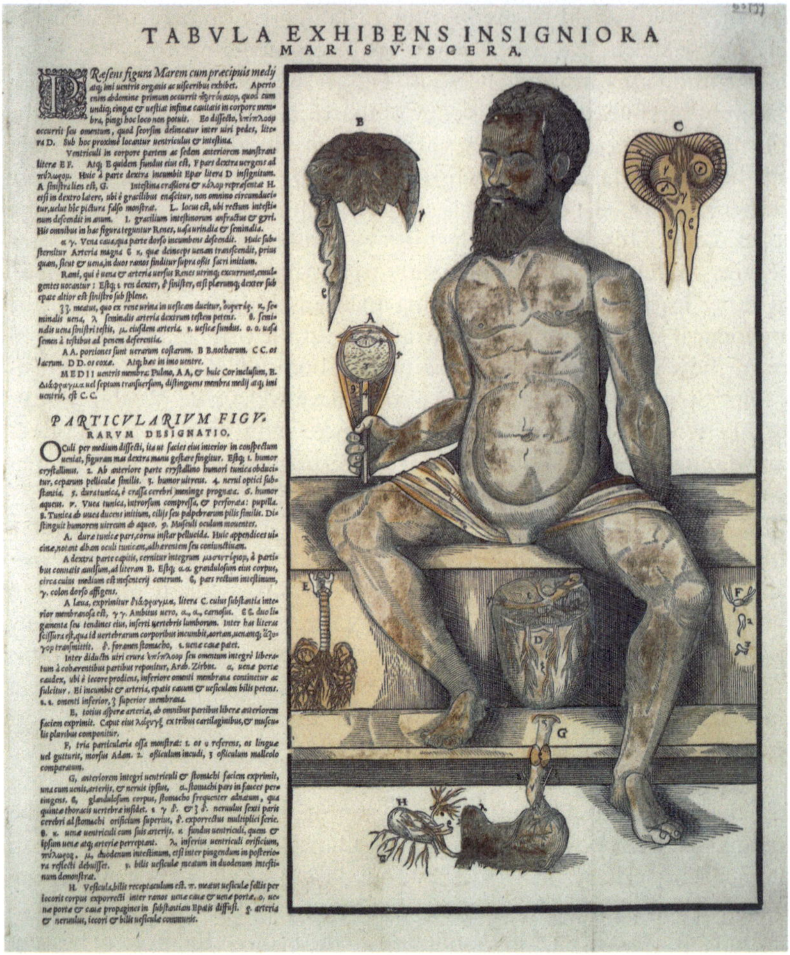

Fig. 4.15 An anatomical fugitive sheet, published at Wittenberg in 1573 to accompany lectures on Melanchthon's treatise on the soul: the head of the male figure clearly parodies that of Vesalius himself. The individual flaps were cut out and stuck down in order, and then raised to reveal the organs lying beneath. (*Source* Wellcome Collection)

the early 1550s which add considerably to our understanding of what was going on there.[223] The learned Cracow physician, Anton Schneeberger (1530–81) not only wrote down many references, some at length, to other anatomists such as Massa and Colombo, but recorded discussions at Montpellier when the dissection was supervised by Rondelet (1507–66). In 1556–7 during the dissection of women there several German students were pleased to discover that the hymen was just as Vesalius had described it, although Rondelet cautioned that its rupture was not itself the cause of pain at first intercourse, but the narrowness of the actual neck of the uterus.[224] The most extensive series of student notes so far discovered in a copy of the *Fabrica* was written by a student attending the lectures of the Granada doctor and surgeon Rodrigo de Molina in the 1550s.[225] He had personally seen Molina's dissections of humans and monkeys, and carefully recorded the many passages where he took issue with Vesalius' criticisms of Galen.

Such detailed annotation becomes much rarer in the seventeenth century, as Vesalius' book becomes less significant as a contemporary manual of anatomy. Maschenbauer's publication (Fig. 3.7), which removed the identifying letters from the plates, as Daniel Margócsy has argued, was a major step in its transition from a practice manual to an object of aesthetic appreciation and connoisseurship.[226] Increasingly also, it was given or bequeathed to an institution, often a Jesuit College or a Protestant University, by one of its members or supporters. When or why the grammar school at Shrewsbury should have come to own a copy by the early seventeenth century baffled its later historian, although he noted that the library also contained several similarly unexpected volumes.[227] This produced a certain stability, especially after the abolition or transformation of the Jesuit Colleges at the end of the eighteenth century and in the aftermath of the French Revolution. By 1850 half of the known copies were now in public or university libraries, and only a third were owned by physicians or surgeons, a reflection of the declining interest in the *Fabrica* as a practical guide to anatomy that is also indicated by a substantial drop in the prices paid between 1750 and 1850.[228] Things began to change at the end of the nineteenth century with the wider growth of medical history first in Europe, especially Germany. The Basle professor of pathological anatomy, Moritz Roth (1839–1914) (Fig. 4.16) published in 1892 the first detailed modern account of Vesalius and the *Fabrica*, which still retains its value today for its medical insights, although two later studies have added much new information.[229] The *Bio-Bibliography* of Vesalius' writings by the distinguished neurosurgeon, Harvey Cushing (1869–1939) (Fig. 4.17), first published posthumously in 1943 and again in a lightly corrected edition in 1962, laid the foundations for the wider study of Vesalius' writings and his influence.[230] It was not, however, until 1964 that Vesalius received a detailed and substantial biography in English and from a scholar trained in history rather than anatomy. The book by C. Donald O'Malley (1907–1970) (Fig. 4.18), professor of the history of medicine at the University of California at Los Angeles, was published in 1964 to coincide with the quincentenary of Vesalius' birth. It is an impressive achievement, even if new discoveries in archives and libraries have since added much and historians have begun to take a more sceptical approach to Vesalius' self-presentation.[231] All three books remain fundamental today and have influenced subsequent historians greatly.

Even more important in the increasing idolisation of Vesalius and his book was the arrival of wealthy American physicians with an interest in

Fig. 4.16 Moritz Roth (1839–1914), Swiss pathologist and author of *Andreas Vesalius* (1892), the first modern biography. (Copyright: The University of Basle)

Fig. 4.17 Harvey W. Cushing (1869–1939), American neurosurgeon and author of *A Bio-Bibliography of Andreas Vesalius* (1943/1962), the standard bibliography of Vesalius and writings associated with him. (*Source* Wellcome Collection)

old and rare books. The Canadian William Osler (1849–1919), who held chairs at Johns Hopkins University in Baltimore and later in Oxford, was a major influence on his colleagues in Baltimore, many of whom purchased their own copies. One 1543 *Fabrica* was even given to a favourite student as a wedding present, and others were passed to institutions like the New York Academy of Medicine.[232] Osler encouraged colleagues across the States to form their own medico-historical libraries and, like him, to donate them to their own universities. In Europe, an American pharmacist, Henry Wellcome, from the 1890s was collecting on a gigantic scale for his Museum and Library.[233] The tendency increased from then on to see the *Fabrica* as a veritable icon of medicine, even as a trophy, for at least one copy was deliberately removed after conquest to the capital of the victors and a legal dispute over ownership of one stolen copy is still unresolved[234] (Fig. 4.19). It has become a sought-after possession for university, state and national libraries, as well as for billionaires, and its price has risen astronomically.[235] If around 1900 one might have paid, in today's money, in the four or low five thousands of pounds for a

Fig. 4.18 C. Donald O'Malley (1907–1970), American historian and author of *Andreas Vesalius of Brussels* (1964), the most detailed biography of Vesalius so far produced. (Copyright: Lance Agnew)

first edition, one would now have to pay nearly twenty or thirty times more, and special editions, like the presentation copy to Charles V or the 1555 annotated edition, sold at Christie's in 2024, much more.[236] Even a second edition, which twenty years ago would have cost the equivalent of a small family car, may reach six figures.[237] Once in a public institutional library, it can reach an audience beyond the merely medical, particularly as an object of art, while the new Census now allows historians to select copies that appear to them to tell a bigger story. With new translations and easy accessibility through the internet, the *Fabrica* is set to attain a new life in the twenty-first century. It may no longer be an up to date guide to anatomy and dissection, but it still retains its power to demonstrate the remarkable interplay of history, science and art in the understanding of the human body.

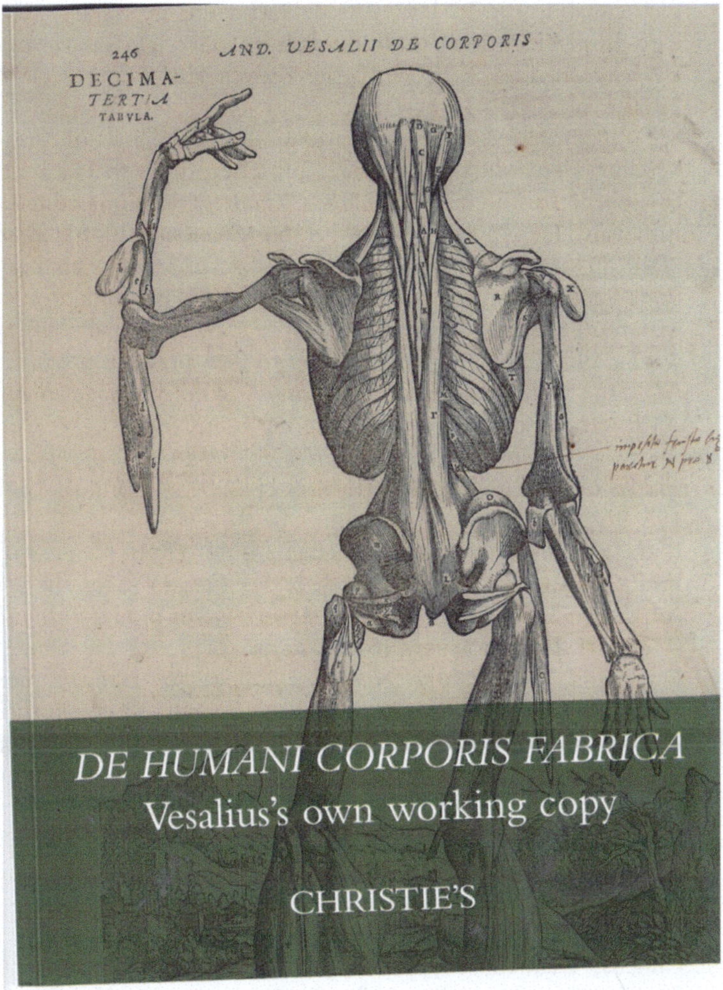

Fig. 4.19 The cover of the *Catalogue* advertising the sale by Christie's on 2 February 2024 of Vesalius' annotated copy of the 1555 edition of the *Fabrica*. The correction to p. 246 is described on p. 179 above. (Copyright: Christie's)

Notes

1. Martin Steinmann, *Johannes Oporinus, ein Basler Buchdrucker um die Mitte des 16. Jahrhunderts* (Basle and Stuttgart: Helbing & Lichtenhahn, 1967), 108; Alfred Hartmann, Beat Jenny, Reinhard Bodenmann, and Lorenz Heiligensetzer, *Die Amerbachkorrespondenz*, 11 vols in 14 (Basle: Universitätsbibliothek, 1942–2010), vol. 8 (1974), xxiv–xxv.
2. See now the letters edited by David Soulier, "The Start of Andreas Vesalius' Career at the Imperial Court According to His Letters to Benedetto Varchi," in *In the Shadow of Vesalius*, edited by Rob Van Hee (Antwerp and Apeldoorn: Garant, 2020), 37–52, 2020, in which he gives some of the background to his hesitations—court intrigue, an offer of a post at Pisa, and problems with the Venetians. There are several examples of papal physicians serving only briefly before returning to their previous posts, Elisa Andretta, Elisa, *Roma medica. Anatomie d'un système médical au XVI*^{e} *siècle* (Rome: École française de Rome, 2011).
3. Andreas Vesalius, *Epistola, rationem modumque propinandi radicis Chynae decocti....pertractans* (Basle: J. Oporinus, 1546), 195–96.
4. Ibid., 136, although Colombo's name is not given.
5. Ibid., 136, with O'Malley, *Andreas Vesalius*, 197; for the male corpse, Ludovico Panizza, *De venae sectione in inflammationibus* (Venice: J. F. Camocius, 1561), 76r.
6. Franciscus Puteus, *Apologia in anatome pro Galeno, contra Andream Vesalium Bruxellensem* (Venice: Francesco di Portinari di Tridino, 1562), 116v–117v, a detailed report mentioning many of those present, possibly including the anatomist Bartolomeo Maggi, cf. Maggi, *De vulneribus sclopetorum et bombardarum curatione* (Bologna: B. Bonardus, 1552), 9v. But see the refutation by Gabriele Cuneo (or Vesalius), *Apologiae Francisci Putei pro Galeno in anatome examen* (Venice: Francesco de' Franceschi, 1564), 45–7, 81, for more details of the dissection.
7. Cuneus, Examen, 40, with O'Malley, *Andreas* Vesalius, 203. Soulier, "Letters to Varchi," 43–44, shows that Vesalius was actively considering a move to Pisa and only finally rejected the idea when he gained a higher salary and an improved position in the imperial medical hierarchy.

8. Tlimann Walter, "Stadt – Hof – Universität. Akademische Ärzte in den Städten und Residenzen des Alten Reichs," in *Personen, Wissen, Karrieren. Bildung und Professionalisierung zwischen Stadt und Hof (1470–1540/50)*, edited by Gerhard Fouquet, Matthias Meinhardt, Sven Rabeler, and Rainer Christoph Schwinges (Ostfilder: Thorbecke, 2024), 181–85.
9. Soulier, "Letters to Varchi," 42–50. They were prepared to see him go to join the Emperor, at least after a year, but not to move to Pisa, whom they suspected of poaching.
10. José Martínez Millán, *La Corte de Carlos V, tercera parte, Los Servidores de las Casas Reale* (Madrid: Sociedad Estatal para la Conmemoración de los Centenarios de Felipe II y Carlos V, 2000), vol. IV, 388; V, 244 (originally on the same salary as his superior, Cornelius Baersdorp) and 247 (Anne, from 1551–56).
11. Soulier, "Letters to Varchi," 43–44; see also the many marginalia to Vesalius' editions of the *Paraphrase*, above, pp. 52–53.
12. Nutton 2019 (originally 1990): 8–11; Walter, "Stadt-Hof-Universität." Andretta, *Mileu médical, passim*, looks at the Papal Court; Richard J. W. Evans, *Rudolf II and His World. A Study in Intellectual History* 1576–1612 (Oxford: Clarendon Press, 1984), 119–24, 203–8, discusses the Imperial court. For erudition as a qualification, Maria Kavvadia, "Sources and Resources of Court Medicine in Mid-Sixteenth Century Rome: Erudition as an Epistemological and Ethical Claim," in *Scientiae in the History of Medicine*, edited by Fabrizio Baldassari and Fabio Zampieri (Rome and Bristol, CT: L'Erma di Bretschneider, 2021), 171–88.
13. Andreas Vesalius, *Epistola, rationem modumque propinandi radicis Chynae decocti....pertractan* (Basle: J. Oporinus, 1546), 197–99. He regretted the loss of connection to the earlier writings by his Vesalian forebears.
14. Ibid., 199, probably misinformed by his brother, ibid., 4.
15. Ibid., 40–41.
16. For his treatment of Don Carlos, the son of Philip II, see Omer Steeno, Maurits Biesbrouck, and Theodore Goddeeris, "Six Previously Unknown Letters from Andreas Vesalius to Octavius Landus Concerning the Accident of the Spanish Prince Don Carlos in 1562," in *In the Shadow of Vesalius*, edited by Rob Van Hee (Antwerp and Apeldoorn, 2020), 53–92.
17. O'Malley, *Andreas Vesalius*, 203.

18. Niccolò Massa, *Epistolae medicinales et philosophicae* (Venice: F. Bindoni and M. Pasini, 1550), 83–87. Moritz Roth, *Andreas Vesalius Bruxellensis* (Berlin: Georg Reimer, 1892), 10, 183–87, discusses Massa's contributions to anatomy here and in his *Introductio.*
19. Michael Stolberg, "Teaching Anatomy in Post-Vesalian Padua: An Analysis of Student Notes," *The Journal of Medieval and Early Modern Studies* 48, no. 1 (2018), 74.
20. Vivian Nutton, *Johann Guinter and Andreas Vesalius, Principles of Anatomy According to the Opinion of Galen* (London and New York: Routledge, 2017), 28, 56.
21. Vesalius, *Epistola Chynae*, 41–43.
22. Ibid., 3–40, a fragment of the longer tract on drugs that he claims to have destroyed in 1544?
23. Ibid., 40–199. The publication of a private letter was a standard gambit in polemics.
24. Daniel H. Garrison, *Andreas Vesalius*, The China Root Epistle. *A New Translation and Critical* Edition (Cambridge: Cambridge University Press, 2015), xxiv–vi.
25. Vesalius, *Epistola Chynae*, 43–44, 95–96.
26. Ibid., 54, 71–72, 85–87, 93–95, 117, 124, 144, 186.
27. Ibid., 136–42, 173–77.
28. Alessandro Benedetti, *De re medica opus insigne apprime Medicinae candidatis omnibus utile* (Basle: Heinrich Petri, 1549). The volume is given over almost entirely to practical medicine, and the treatise on anatomy takes up only 47 of its 623 folio pages.
29. Ibid., sigg. A 2r-v. David Soulier, *Vésale anatomisé. Réception et diffusion du* De humani corporis fabrica *en Europe (1543–1628)* (PhD. Diss. Université Côte d'Azur, 2020), 195–97. The whole thesis offers a panoramic view of contemporary reactions to Vesalius.
30. Frank Hieronymus, *Theophrast und Galen—Celsus und Paracelsus. Medizin, Naturphilosophie und Kirchenreform im Basler Buchdruck bis zum dreissigjährigen Krieg* (Basle: Universitätsbibliothek, 2005), 189–93; Giovanna Ferrari, *L'esperienza del passato. Alessandro Benedetti, filologo e medico umanista* (Florence: L. S. Olschki, 1996), 326. This is rather different from Fuchs' giving additional references to the *Fabrica* in the

margins of his 1551–5 Epitome of the Fabrica, signalled to me by Monique Kornell.

31. For Spain, Bjørn O. Skaarup, *Anatomy and Anatomists in Early Modern Spain* (Farnham and Burlington VT: Ashgate, 2015); Carlos Fernando Teixeira Alves, "The Teaching of Anatomy the South of Europe: The Case Study of the Universities of Coimbra and Salamanca (XIV-XVI)," in *The Spaces of Renaissance Anatomy Theater*, edited by Leslie R. Malland (Wilmington and Malaga: Vernon Press), 45–68; for Colombo, Realdo Colombo, *De re anatomica libri XV; Anatomia. Edited with translation and commentary by G. Baldo* (Paris: Les Belles Lettres, 2014), LXXII–IX.
32. Vivian Nutton, "The Anatomy of the Soul in Early Renaissance Medicine," in *The Human Embryo. Aristotle and the Arabic and European Traditions*, edited by Gordon R. Dunstan (Exeter: University of Exeter Press, 1990), 148–49; idem, "Wittenberg Anatomy," in *Medicine and the Reformation*, edited by Ole P. Grell and Andrew Cunningham (London and New York: Routledge, 1993), 11–32; David Cellamare, "History of a Productive Failure. The Use of Andreas Vesalius's *De humani corporis fabrica* in some Sixteenth-century Books on the Soul," in *Towards the Authority of Vesalius: Studies on Medicine and the Human Body from Antiquity to the Renaissance and Beyond*, edited by Erika Gielen and Michèle Goyens (Turnhout: Brepols, 2018), 403–36; for a Lutheran theologian recommending the study of anatomy to his pupils at Rostock, above, p. 208. See also Fig. 4.15.
33. Leonhart Fuchs, *De humani corporis fabrica epitomes….pars prima* (Lyon: J. Frellon, 1551); *De humani corporis fabrica epitome….pars altera* (Lyon: J. Frellon, 1555).
34. Vesalius thought highly of his work as an anatomist and would have liked him to have taken the chair of anatomy at Pisa that he himself had been offered, Soulier, "Letters to Varchi," 44–50.
35. Jacobus Sylvius, *Vaesani cujusdam calumniarum in Hippocratis Galenique anatomicam depulsio* (Paris: C. Barbé and the widow of I. Gazellus, 1551), 29r.
36. Renatus Henerus, *Adversus Jacobi Sylvii depulsionum anatomicarum calumniarum, pro Andrea Vesalio apologia* (Venice: n.p., 1555), 1–5. Soulier, *Vésale anatomisé*, 286–94 argues that Vesalius was here using the name of Hener as a disguise, for which

he rewarded Hener with a copy of the *Fabrica* (which he must not have owned earlier). The note by his son on this copy, Daniel Margócsy, Mark Somos and Stephen N Joffe, *The* Fabrica *of Andreas Vesalius. A Worldwide Descriptive Census, Ownership and Annotations of the 1543 and 1555 Editions* (Leiden and Boston: Brill, 2018), 311–12, shows that at least the family believed that Hener was the author.

37. Franciscus Puteus. *Apologia*, 7-10r, 13r; 17v. For the controversy with Cuneo, Arturo Castiglioni, "The Attack of Franciscus Puteus on Andreas Vesalius and the defence by Gabriel Cuneus," *Yale Journal of Biology and Medicine* 16 (1943), 135–42; Soulier, *Vésale anatomisé*, 443–58.
38. Cuneus, Examen; Soulier, *Vésale anatomisé*, 451–57, following much earlier suggestions by Cardano and Boerhaave and Albinus, argues strongly that this was written by Vesalius himself under the pseudonym of Cuneo because the author shows a detailed knowledge of the unpublished *Examen* as well as of the Bologna disputation of 1544, Cuneus, *Examen*, 17–18, 81.
39. Skaarup, *Early Modern Spain.*
40. Stolberg, "Teaching Anatomy," 73–74.
41. Howard M. Adelmann, *Marcello Malpighi and the Evolution of Embryology* (Ithaca: Cornell University Press, 1966), 86–87.
42. Cynthia Klestinec, *Theaters of Anatomy: Students, Teachers and Traditions of Dissections in Renaissance Italy* (Baltimore: Johns Hopkins University Press, 2011).
43. Klestinec, *Theaters*, 19–35, 42–43.
44. Jacopo Facciolati, *Fasti gymnasii Patavini ab anno mdxvii quo restitutae scholae sunt ad mdcclvi* (Padua: Giovanni Manfrè, 1757), 386, noting that Vesalius' temporary resumption of his old duties attracted a huge audience.
45. Sigismund Gelenius, *Notitia utraque cum orientis tum occidentis ultra Arcadii Honoriique caesarum tempora* (Basle: Hieronymus Froben and Nicolaus Episcopius, 1552), pref.
46. Andreas Burchkhardt, *Johannes Basilius Herold. Kaiser und Reich im protestantischen Schrifttum des Basler Buchdrucks um die Mitte des 16. Jahrhunderts* (Basle and Stuttgart: Helbing & Lichtenhahn, 1967), 104.
47. O'Malley, *Andreas Vesalius*, 263.

48. Michael Horowitz and Jack Collins, "A Census of Copies of the First Editions of Andreas Vesalius' *De humani corporis fabrica* (1543) with a Note on the Recently Discovered Issue," *Journal of the History of Medicine and Allied Sciences* 39 (1984), 198–221. The changes are more considerable than the simple corrections made during the actual printing process found in some copies, Stephanie Charreaux and Jérôme Van Wijland, "Recensement et déscription des exemplaires de la première édition du De Fabrica (1543) conservés en France dans les bibliothèques publiques," in *La* Fabrique *de Vesale. La mémoire d'un livre*, edited by Jacqueline Vons (Paris: Bibliothèque interuniversitaire de Santé, 2016), 253–312.
49. Above, p. 118.
50. Above, pp. 162–23. But some copies were still available, along with the *Epitome*, in early 1547, O'Malley, Andreas Vesalius, 228. For sales in Italy, see Vesalius' letter of 1544 to Benedetto Varchi, Soulier, "Letters to Varchi," 50–51.
51. Margócsy et al., *The Fabrica*, 237–38; Vivian Nutton, "Vesalius in England, 1544 1547," in *In the Shadow of Vesalius*, edited by Rob Van Hee (Antwerp and Apeldoorn: Garant, 2020), 93–100; for the later use of Vesalius by Thomas Lorkyn in Cambridge, Peter M. Jones, "Thomas Lorkyn's Dissections, 1564/5 and 1566/7," *Transactions of the Cambridge Bibliographical Society* 9, no. 3 (1988), 209–29.
52. Christopher Langton, *An Introduction into Physicke, with an Universal Dyet* (London: E(dward) W(hitchurch), undated), chs 5–6.
53. Christopher Langton, *A very brefe treatise ordrely declaring the principal partes of physick, that is to say: - Thynges naturall, Thynges not naturall, Thynges against* nature (London: Edward Whitchurch, 1547), sig. d 4 r. Spelling modernised.
54. William Munk, *The Roll of the Royal College of Physicians of London: 1518 to* 1700 (London: Royal College of Physicians, 1878), 151–52; Nutton, "Vesalius in England."
55. Melissa Lo, "Cut, Copy and English Anatomy: Thomas Geminus and the Reordering of Vesalius's Canonical Body," in *Andreas Vesalius and the "Fabrica" in the Age of Printing: Art, Anatomy, and Printing in the Italian Renaissance*, edited by Rinaldo

Canalis and Massimo Ciavolella (Turnhout: Brepols, 2018), 225–53.

56. Thomas Geminus, *Compendiosa totius anatomiae delineatio aere exarata* (London: J. Herford, 1545), pref.; ed. 2, *Compendiosa totius anatomiae delineatio aere exarate* (London: N. Hyll for T. Geminus 1553); Iain M. L Donaldson, "Two States of Some Plates in the *Compendiosa* of Thomas Geminus (1545)," *The Library* 11, no. 1 (2010), 89–104.
57. Vesalius, *Epistola Chynae,* 198.
58. Vesalius, *Anatomia deudsch; ein kurzercAuszug der Beschreibung aller Glider menschlichs Leybs* (Nuremberg: J. P. Fabricius, 1551); Geminus, *Delineatio*, 1553; Harvey W. Cushing, *A Bio-Bibliography of Andreas Vesalius* (Hamden, CT, and London: Archon Books, 1962), 121–53, discusses Geminus and other works descending from him (cf. also Elly Cockx-Indestege, *Andreas Vesalius, A Belgian Census: Contribution Towards a New Edition of H. W. Cushing's* Bibliography (Brussels: Royal Library,1994), 88–122).
59. Above, pp. 24–25, 52–53.
60. For the fonts, borrowed from Enzinas, see above, pp. 168–69.
61. O'Malley, *Andreas Vesalius*, 463, n. 6.
62. Ibid., 270. Was the reference to a new and improved edition an attempt to test the water for sales of a future edition?
63. Steinmann, *Oporinus*, 37, 87–88.
64. O'Malley, *Andreas Vesalius*, 270, but Oporinus may simply have been expecting to have the complete text to hand within an extremely short time, which, given Vesalius' propensity to correct and emend frequently and quickly from the moment of the original printing, is not impossible.
65. Was it made to cover the sheer costs of the paper, which, according to Philip Gaskell, *A New Introduction to Bibliography* (Winchester: St Paul's Bibliographies, 1995), 177, might have amounted to two-thirds of the expenses of a publisher.
66. In 1561 Herold claimed that he had lost 600 guilders in subsidising the work, Burckhardt, "Herold," 189; on this publication and its problems, ibid., 173–93.
67. Ibid.; Carlos Gilly, *Die Manuskripte in der Bibliothek des Johannes Oporinus* (Basle: Schwabe, 2001), 13. Theodore Zwinger, in a letter printed by Soulier, *Vésale anatomisé*, 585, gives the sum

as roughly 2000 florins. See also below, p. For Vandermeeren in perhaps 1578, Hartmann et al., *Amerbachkorrespondenz*, vol. 11 (2010), 1011, undated, but certainly after 1575. For the possibility that this was a subsidy towards a third edition, see above, p. 168.

68. O'Malley, *Andreas Vesalius*, 270.
69. The rest of this paragraph depends on Carlos Gilly, *Spanien und der Basler Buchdruck bis 1600: Ein Querschnitt durch die spanische Geistesgeschichte aus der Sicht einer europäischen Buchdruckerstadt* (Basle: Helbing und Liechtenhahn, 1985), 327–28, 333–40. Enzinas is also known by his Latinised name Dryander, not to be confused with Vesalius' Marburg anatomist of the same name.
70. This means that Oporinus must have cast the fonts he planned to use for the first five books and returned the matrices to Enzinas in early 1550.
71. O'Malley, *Andreas Vesalius*, 270.
72. Roger Gaskell, "Typography," 45–47.
73. Above, p. 179 and; Fig. 4.19; O'Malley, *Andreas Vesalius*, 463–64, n. 20.
74. Barry J. Anson, "The Initial Letters of the 1555 Edition of Vesalius's *Fabrica*," *Quarterly Bulletin of the Northwestern University Medical School* 19 (1943), 326–35 [p. 330]; Adolf Faller, *As iniciais da II^{a} ediçião (1555) de obra 'De humani corporis fabrica Lib;ri septem' da Andrea Vesalio* (Rio de Janeiro: CIBA, 1951); Frank Hieronymus, *Oberrheinische Buchillustration 2, Basler Buchillustration 1500–1545* (Basle: Universitätsbibliothek, 1984), 648–49; Guy Gilias, "Ornamental letters in the Fabrica 1543 and 1555," in *Art of Vesalius*, edited by Rob Van Hee (Antwerp and Apeldoorn: Garant, 2014), 97–112.
75. Vesalius, *Fabrica* 1543, 228/328 = 1555, 396. Above, p. 118.
76. The most detailed study of the 1555 initials is that of Faller, "As inicials," known to me through a copy of an offprint from the Lausanne Medical Library, kindly provided to me by Omer Steeno and Maurits Biesbrouck.
77. The initials E, F, I, N and S reverse the original design, but this need not indicate the activity of two different designers or cutters.
78. Garrison and Hast, *The Fabric of the Body*, translate all the changes in this edition.

79. O'Malley's observation, *Andreas Vesalius*, 272, that some verbal changes would have involved the cutting of a new initial is almost irrelevant since all the initials in the chapter headings bar one were recut, whereas only one of the opening Book initials, V, had to be newly cut.
80. For these initials, see above, p. 101.
81. O'Malley, *Andreas Vesalius*, 274–75.
82. Vesalius, *Fabrica* 1543, 19, 266.
83. Ibid., 1543, 27, 159–62.
84. Ibid., 1543, 538; 1555, 662.
85. Ibid., 1543, 584; 1555, 728.
86. Ibid., 1543, 585, 633–34.
87. Vesalius, *Fabrica* 1555, 734, 736.
88. Ibid., 1555, 669–70, 588.
89. Vesalius, *Fabrica* 1543, 259; 1555, 307: 17; *Fabrica* 1543, 295/395; 1555, 483. O'Malley, *Andreas Vesalius*, 463–64, n. 20, lists further changes to the plates.
90. Vesalius, *Fabrica* 1543, 262/362, 335/435; 1555, 444, 530.
91. Above, p. 192.
92. Vesalius, *Fabrica* 1555, 442–43.
93. Ibid., 595, 589.
94. Ibid., 540; 442–43, 654; 23–24.
95. Erik Waller, "Eine unbekannte Ausgabe von Vesals *Epitome*," *Lychnos* (1936), 152–60; Cushing, *Bio-bibliography*, 115–16.
96. David S. Chambers, "A Mantuan in London in 1557: Further Research on Annibale Litolfi," in *England and the Continental Renaissance. Essays in Honour of J. B. Trapp*, edited by Edward Chaney and Peter Mack (Woodbridge: The Boydell Press, 1990), 96.
97. Andreas Vesalius, *Anatomicarum Gabrielis Falloppii observationum examen* (Venice: Francesco de' Franceschi, 1564), 171. O'Malley, *Andreas Vesalius*, 285–88, covers the death of Henri II in detail. For Spanish Vesalians, Skaarup, *Early Modern Spain*, 25–28; Teixeira Alves, "Teaching of Anatomy," 48–57.
98. Above, pp. 176–87. The identification was first made by Gerard Vogringic in late 2007, who then asked me to assist him. The volume was sold at Christie's on line on February 2, 2024, Christie's, *De humani corporis fabrica. Vesalius' own Working Copy*, Sale Catalogue, online sale, January 17 to February 2,

2024. References in the rest of this chapter to the 1555 edition are to this copy.
99. The most significant changes are translated into English in the notes to Garrison and Hast, *The Fabric of the Body*.
100. This section is based on Vivian Nutton, "Vesalius Revised: His Annotations to the 1555 *Fabrica*," *Medical History* 56 (2012), 415–43.
101. Steinmann, *Oporinus*, 38–39, 47–48, calculates that the index would have taken between eight to ten days to set, and the Errata a matter of hours.
102. Its provenance before it was sold by its German owner in 2007 at auction in Hamburg, is unclear. The circular indentation on the spine once contained a paper circle with perhaps a classmark, but this fell off when the volume arrived in Canada from Germany, so its later owner told me, and is now lost. This would further suggest that it had once been in an institutional library.
103. See Fig. 4.19.
104. For contemporary discussion of this muscle, see Stolberg, *Falloppia*, 53.
105. A similar comment about Celsus is made on p. 135, line 2, referring to Celsus, *De medicina* III,1,19–20; 10,2–3.
106. Hippocrates, *Fabrica* 1555: 18,9; 124,6; 181,28; Celsus, 129,1; 135,7; Galen, 516, bottom; Oribasius, 340.11 from end (and then deleted); Paul, 640,20; Cicero, 58.14.
107. Ibid., 72,5 and 207,23–24; 207,23–24, with Robert Herrlinger and Edith Feiner, "Why Did Vesalius Not Discover the Fallopian Tubes?" *Medical History* 8 (1964), 335–41. He does not appear to have heard of Falloppia's discoveries described in the *Observations*.
108. Vesalius, *Fabrica 1555:* 481,13; 721,24.
109. Ibid., 653, bottom.
110. Damian De Goes, *Fides, mores, religioque* Aethiopum (Louvain: R. Rescius, 1549), 70.
111. Paolo Giovio, *Historiae sui temporis* (Florence: L. Torrentino, 1550), 304.
112. Nutton, "Vesalius Revised," 434. gives a sample list.
113. Vesalius, *Fabrica* 1555, 65,14, information from Professor Iain Donaldson.

114. Realdo Colombo, *De re anatomica* (Venice: N. Bevilacqua, 1559), 219; Tawrin Baker, "Why All This Jelly? Jacopo Zabarella and Hieronymus Fabricius ab Aquapendente on the Usefulness of the Vitreous Humour," in *Early Modern Medicine and Natural Philosophy*, edited by Peter Distelzweig, Benjamin Goldberg, and Evan R. Ragland (Wiesbaden: Springer, 2016), 59–88. Both Berengario and Benedetti had earlier rejected this mediaeval placing of the lens in the centre of the eye, a comment I owe to Tawrin Baker, and it was correctly shown also by Colombo's student Valverde in 1556, Juan Valverde de Hamusco, *Historia de la composicion del cuerpo humano* (Rome: A. Salamanca and A. Lafreri, 1556), Book 5, plate 3, information from Monique Kornell.
115. He repeated his conclusion in the *Examen*, 162.
116. Jacqueline Vons and Stéphane Velut, *André Vésale. Résumé de ses livres sur la Fabrique du corps* humain (Paris: Les Belles Lettres, 2008), XXXIII.
117. Andreas Vesalius. *De humani corporis fabrica* (Venice: Francesco de' Franceschi, 1568), sig. *6v; see above, p. 195.
118. For example, he omits the corrections in the Errata on p. 9,32; p. 17, caption, line 12; p. 632, caption, line 3; on sig. A 2v,20 he prints *familiae* for *familia*, and on p. 573,3 *relicto* instead of *relicta*. He does not make the obvious transposition at p. 155,15–20, and the captions on p. 243 are headed *DUOCECIMAE*.
119. Gabriele Falloppia, *Observationes anatomicae* (Venice: Marcantonio Olmo and Gratioso Perchachino, 1561). Stolberg, *Falloppia*, is the best modern study.
120. Some of Falloppia's views he corrects, and others he passes over in silence, see now the detailed edition of the *Examen* by David Soulier, *Examen des* Observations anatomiques *de Gabriel Fallope* (Paris: Classiques Garnier, 2023).
121. Colombo, *De re anatomica*, 216–17; Vesalius, *Examen*, 47–48; above, p. 193. But he may not be referring to Colombo but to his Roman rival, Bartolommeo Eustachi, for whose critical attitude to Vesalius, see Andretta, *Roma medica*, 528–30. For Falloppia's reactions, above, p. 192.
122. Vesalius, *Examen*, 24; Albert Mudry, "Disputes Surrounding the Discovery of the Stapes in the Mid-16th Century," *Otology & Neurotology* 34, no. 3 (2013), 588–92. Vesalius leaves it unclear

whether he had made the same discovery before or after reading Falloppia.

123. A suggestion made to me by Martin Steinmann.
124. Above, p. 168. Or was the 2582 Gulden a loan to pay for the third edition? The dispute continued after 1575.
125. Above, pp. 23–24, 52.
126. Vesalius, *Fabrica* 1555, 805,6.
127. Ibid., 196,3 (advice on instruments); 287,6 from end (how to cut into a skull and remove eye-muscles); 352,2–5 (questions to be asked after freeing the muscles of the thorax).
128. Rinaldo F. Canalis, "Gabrielle Falloppia; Vesalius's Admirer and First Critic," in *Andreas Vesalius and the "Fabrica" in the Age of Printing: Art, Anatomy, and Printing in the Italian Renaissance*, edited by Rinaldo F. Canalis and Massimo Ciavolella (Turnhout: Brepols, 2018), 171–200; Stolberg, Falloppia, 30–35. He had previously taught at Ferrara and from 1558 to 1551 in Pisa, where he made his reputation.
129. Falloppia, *Observationes*, e.g. 1r-v, 25r, 31r, 34r, 39r, 222r-v; 92v, "divine Vesalius" as a miracle; the book itself, 2r-v; at 73r Falloppia talks of things missed by anonymous "divine anatomists" (including Vesalius).
130. Ibid., 2v-3r.
131. Pietrandrea Mattioli, *Epistolae medicinales* (Venice: Vincenzo Valgrisi, 1561), 151–52.
132. Falloppia, *Observationes*, 25v-26r; above, p. 185; 26v, for communicating by letter to fellow anatomists. Cf. ibid., 102v-103r, on Canani's discovery of the *palmaris brevis* muscle, doubted by Vesalius in 1542 and later. For other early improvements by Ingrassia, cf. Gian Filippo Ingrassia, *De tumoribus praeter naturam* (Naples: M. Cancer, 1553), 156–58.
133. Falloppia, *Observationes*, 6r-v, with the overall aim of defending the doctrines of Galen.
134. His discoveries are clearly set out by Stolberg, *Falloppia*, 45–58.
135. Falloppia, *Observationes*, 21r-32r, 36r, 39r, 46r, 47v, 51r-53v. Cf. 88r for a possible anomaly as an explanation for an error of Vesalius.
136. Ibid., 19r.
137. Ibid., 47r-v.
138. Ibid., 12r and 16r (ginglymus hinge); 53r-54v (deltoid muscle).

139. Ibid., 75r-v; cf. 82r for a similar alleged mistake by Vesalius.
140. Ibid., 118r-19r; cf. Vesalius, *Examen*, 83, and the discussion in Stolberg, *Falloppia*, 55–57.
141. Stolberg, *Falloppia*, 59–60. For Berengario's (and possibly Benivieni's) priority in insisting of the importance of seeing and touching, see above, p. 32; Victoria von Hoffmann, "Epistemologies of Touch in Early Modern Autopsies," *Renaissance Quarterly* 75 (2022), 542–82.
142. Vesalius, *Examen*, sig. +2r-v.
143. Ibid., 2.
144. Ibid., 2, 171 (a hint that he might return to academia?).
145. Ibid., 2, 170–71.
146. Ibid. 71.
147. Ibid., 80 (his disagreement with Corti at the Bologna dissection); 82–83.
148. Katharine Park, "The Rediscovery of the Clitoris," in *The Body in Parts: Fantasies of Corporeality in Early Modern Europe*, edited by David Hillman and Carla Mazzio (London and New York: Routledge, 1997), 176–78.
149. Vesalius, *Examen*, 72–73. Carolina Alarcon, "Imitation is the Sincerest Form of Innovation: Valverde Reconsidered," *The Sixteenth Century Journal* 53, no. 1 (2022), 1–39, stresses the improvements made by him in his illustrations, printed from copper plates and his slightly different pedagogical perspective, as well as Vesalius' self-promotional skills.
150. Vesalius, *Examen*, 92–93, 70 (where he also declares that he had no copy of Galen's book on muscles with him and that there was no chance of finding a copy in Spain), 73.
151. Ibid., 57.
152. Ibid., 33, 24.
153. Ibid., 71, 83.
154. Ibid., 37–38.
155. Ibid., 17–18.
156. Ibid., 15.
157. Ibid., 47–48; Stolberg, *Falloppia*, 53.
158. Soulier, *Examen*, 20–23, a more favourable judgment than that of O'Malley, *Andreas Vesalius*, 289–96.

159. For Landi's murder, see Stolberg, *Falloppia*, 152–63; Fabrizio Bigotti, "Commenting on Aristotle with a Knife. The Heretical Anatomies of Bassiano Landi," *NTM, n.s.* 31 (2023), 1–26.
160. By now the authorities had reverted to using a surgeon to do the actual cutting, and Lendinara had acted in this role for the ailing Falloppia.
161. Georg Purkircher, *Opera quae supersunt omnia, edidit Miloslaus Okál* (Budapest: Akadémiai Kiadó.1988), 153, 155, 156. The deadly rivalry between Poles and Germans did not help; the date on the first letter is February 18, 1563, but this cannot be right as it refers to events in the following month, and 18 March is more likely.
162. Above, p. 193; Theo Dirix, *In Search of Andreas Vesalius. The Quest for the Lost Grave* (Tielt: Lannoo Campus, 2014), 112.
163. Pietro Bizzarri, *Historia della guerra fatta in Ungheria* (Lyon: n.p., 1568), 178. For Bizzarri's career as a religious exile and possibly a spy, Silvana Menchi, "Bizzarri, Pietro," in *Dizionario biografico degli italiani* (Rome: Treccani, vol. 10 (1968)), 38–41.
164. Vesalius, *Examen*, 171.
165. For similar (and perhaps better justified) stories of Falloppia, see Stolberg, *Falloppia*, 37–42.
166. Contemporary estimates of the *amount* he left to his wife at his death put it at an income of 5,000 *livres* and capital of 30,000 *livres* or at the somewhat more modest 3,000 florins and 10,000 florins of capital. His Brussels house was valued for tax purposes in 1571 at 10,000 florins, O'Malley, *Andreaas Vesalius* 1964: 313.
167. Dirix, *In Search of Vesalius*, reviews the theories about his death.
168. Above, p. 90.
169. Above, p. 99.
170. Above, p. 186.
171. Felix Platter, *De corporis humani structura et usu* (Basle: A. Froben, 1583), pf.; Frank Hieronymus, *Oberrheinische Buchillustration 2, Basler Buchillustration 1500-1545* (Basle: Universitätsbibliothek, 1984), 650–51 (=. *Theophrast und Galen*, 3035–51, with more images) prints the Latin text of Platter and lists other smaller copies of the images.
172. In June 1557 Platter asked Oporinus for a (free?) copy of the 1555 edition, Alfred Hartmann et al., *Die Amerbachkorrespondenz*, vol. 10, 699.

173. Andreas Maschenbauer, *Andreae Vesalii Bruxellensis, Deß Ersten, Besten Anatomici, Zergliederung Deß Menschlichen Körpers. Auf Mahlerey und Bildhauer-Kunst gericht* (Augsburg: Andreas Maschenbauer, 1706). Quotations from unpaginated preface and letter to the reader. For Tortebat and Bonaveri see above, p. 86.
174. John A. Goree, "Woodblocks," *Printing History* 16 (2014), 3–41, is the best modern study of the history of the blocks. Willy Wiegand, "Marginal Notes," by the Printer of the ICONES, in Lambert et al., *Three Vesalian Essays*, 27–42, remains valuable. See also Daniel Margócsy, "From Vesalius through Ivins to Latour: Imitation, Emulation and Exactly Repeatable Pictorial Statements in the *Fabrica*," *Word & Image* 35, no. 3 (2019), 315–33.
175. Heinrich Palmaz Leveling, *Anatomische Erklärung der Original-Figuren von Andreas Vesal: Samt einer Anwendung der Winslowischen Zergliederungslehre in sieben Büchern*. Ingolstadt: Anton Attenkhouer Leveling 1783 fol. xx2r, citing the 1754 German edition of Winslow's popular book of anatomy, *Abhandlung von dem Bau und der Zergliederung des menschlichen Leibes* (Basle, 1754).
176. Leveling, *Anatomische Erklärung*; the preface, with the details given here, is translated in Cushing, *Bio-bibliography*, 101–4.
177. Moritz Roth, "Vesaliana," *Virchows Archiv* 141 (1895), 462–78.
178. Wiegand, "Marginal Notes," tells his own story; see also Iain M. L. Donaldson, "Andreae Vesalii Bruxellensis Icones anatomicae. Part 2," *The Journal of the Royal College of Physicians of Edinburgh* 42, no. 3 (2012), 280–82; Goree, "Woodblocks", 26–38.
179. Andreas Vesalius. *Andreas Vesalii Bruxellensis "Icones anatomica,"* (Munich and New York: Library of the University of Munich and the New York Academy of Medicine, 1935).
180. Goree, "Woodblocks," 39–41.
181. Caspar Bauhin, *De corporis humani fabrica: libri IIII. Methodo anatomica in praelectionibus publicis proposita, ad Andreae Vesalii tabulas instituta: sectionibusque publicis et privatis, comprobata* (Basle: S. Henricpetri, 1590).
182. For the marquetry panels, Deanna Petherbridge, *The Quick and the Dead. Artists and* Anatomy (London: The South Bank Centre, 1997), 54; for a tomb slab of 1716 in Valletta, Dane Munro, *Memento Mori. A Companion to the Most Beautiful Floor in the World* (Valletta: MJ Publications, 2005), 303,

no. 198; a later tomb slab of 1832 has two Vesalian skeletons, http://altereia.org/2017/05/24/walking-on-tombstones-in-aMaltese-church, consulted on line on 9 February 2024.

183. Clément Baudin, *Description et demonstration des membres interieurs de l'homme et de la femme ....selon la vraye anatomie de André Wesale....Oeuvre utile & nécessaire non seulement aux Medecins & Chirurgiens, ains aussi aux Portrayeurs & Architectes* (Lyon: C. Baudin, 1560); Kornell, *Flesh and Bones*, 30–31; Monique N. Kornell and Daniel Margócsy, "'A Spring of Immortal Colours': Jacques Le Moyne de Morgues (c. 1533–1588) and Picturing Plants in the Sixteenth Century," *Journal of the Warburg and Courtauld Institutes* 86 (2023), 109–57.
184. Hans L. Houtzager, "Editions, Copies and Replicas of Fabrica and Epitome, Published Between 1543 and 1725," in *Art of Vesalius*, edited by Rob Van Hee (Antwerp and Apeldoorn: Garant, 2014), 190–220, discusses some 'variations' on the *Fabrica* and its *Epitome.*
185. Saunders and O'Malley, *Illustrations.* I I cite it from the 1982 reprint, with a slightly different title. *The Anatomical Drawings of Andreas Vesalius* (New York: Bonanza Books), which reduces the size of some images.
186. Andreas Vesalius, *De humani corporis fabrica* (Lyon: Jean de Tournes, 1552), pref.; Cushing, *Bio-bibliography*, 88–90.
187. Cushing, *Bio-bibliography*, 18–53, lists publications that use only the images either entire or in part; for Valverde, ibid., 146–53 (from 1556 onwards in various languages), with Bjørn O. Skaarup, "The Unexpected Success of a Spanish Anatomy Book: Juan Valverde de Amusco's *Historia de la composicion del cuerpo humano* (Rome, 1556), and Its Many Later Editions," in *Specialist Markets in the Early Modern Book World*, edited by Richard Kirwan and Sophie Mullins (Leiden: Brill, 2015), 123–41.
188. Leonhart Fuchs, *De humani corporis fabrica epitome* (Lyon: J. Frellon, 1551–55); Cushing, Bio-bibliography, 136–37; Soulier, *Vésale anatomisé*, 230–34; above, p. 163. Volume 1 was quickly reprinted at Lyon by Antoine Vincent, in 1551, volume 2 in 1555.
189. Andreas Vesalius, *De humani corporis fabrica* 1568: Cushing, Bibliography, 92–93; Soulier, *Vésale anatomisé*, 230–34; 584–85.

190. Andreas Vesalius, *Anatomia* (Venice: Giovanni Antonio and Giacomo de' Franceschi, 1604). In the dedication to the Duke of Urbino they claim republished the book because copies were no longer available in bookshops.
191. Vesalius, *Anatomia*: Cushing, *Bio-bibliography*, 93–94.
192. Vesalius, *Anatomia*, vol. II, sog. A 2r.
193. Carlo Ruini, *Anatomia del cavallo, infermità, e suoi rimedii* (Bologna: Eredi di Giovanni Rossi, 1598): Peter E. J. Bols, "How Andreas Vesalius Inspired Early Veterinary Authors," in *In the Shadow of Vesalius*, edited by Rob Van Hee (Antwerp and Apeldoorn: Garant, 2020), 256–59.
194. Cushing, *Bio-bibliography*, 220–22.
195. Vesalius, *Opera omnia Anatomica et Chirurgica* (Leiden: J. de Vivie and J. and H. Verbeek, 1725); Cushing, *Bio-bibliography*, 133–34. A preliminary printing of the first three books appeared in 1724, presumably to attract prospective purchasers and subscribers for the full publication (which engaged a total of 188 subscribers, mainly from Holland).
196. Ian Maclean, "Vesalius as Authority: From the First Publications to the *Opera Omnia* of 1725," in *Towards the Authority of Vesalius: Studies on Medicine and the Human Body from Antiquity to the Renaissance and Beyond*, edited by Erika Gielen and Michèle Goyens (Turnhout: Brepols, 2018), 42, calls it 'most plausibly a case of vanity purchasing perhaps prompted also by national pride.'
197. Soulier, *Vesale anatomisé*, 819–22, with a translation of the preface at 1259–86; cf. also Hélène Cazes, "Réédition et retour au progrès: les Œuvres d'André Vésale (Leiden, Boerhaave et Albinus, 1725), acte de naissance et de renaissance de l'anatomie," in *La fabrique de la modernité scientifique. Discours et récits du progrès sous l'Ancien Régime*, edited by François Charbonneau (Oxford: Voltaire Foundation, 2015), 9–35.
198. Published respectively in Brussels: Culture et Civilisation, 1964/1970; [Nieuwendijk N.B.]: [De Forel], [1975].
199. Andreas Vesalius, *André Vésale, La Structure du corps humain*, translated by Jules-Émile Verschaffelt (Haarlem: Enschede & Sons, 1924); *Vesalius on the Human Brain*, translated by Charles Singer (London: Oxford University Press for the Wellcome

Historical Medical Museum, 1952); other 'fragments' are listed in Cushing, *Bo-bibliography*, 95–96.

200. Vesalius, *De humani corporis fabrica*, translated by Vasiliy N. Ternovsky and Sergei P. Shestakov (Moscow: USSR Academy of Sciences, 1950–54). I am grateful to Lena Berger for information on this edition.
201. Andreas Vesalius. *Vesalius, On the Fabric of the Human Body*, translated by William F. Richardson and John B. Carman (Novato, CA: Norman Publishing, 1998–2009).
202. Andreas Vesalius, *Andreas Vesalius, The Fabric of the Human Body*, translated by Daniel H. Garrison and Malcolm H. Hast (Basle: Karger, 2014).
203. The same website also provides an easily accessible and almost complete listing of editions of the works of Vesalius. A CD Rom of the 1543 edition, with supplementary material, was produced in 2003 at Palo Alto.
204. Elly Cockx-Indestege, *Andreas Vesalius, A Belgian Census: Contribution Towards a New Edition of H. W. Cushing's* Bibliography (Brussels: Royal Library, 1994).
205. Daniel Margócsy, Mark Somos, and Stephen N. Joffe, *The Fabrica of Andreas Vesalius. A Worldwide Descriptive Census, Ownership and Annotations of the 1543 and 1555 Editions* (Leiden and Boston: Brill, 2018); Soulier, *Vésale anatomisé*, 773–80.
206. I know of one unregistered copy of the 1555 edition in private hands, and another was sold at Christie's on October 9, 2020. Copy I/P13 is now in the New South Wales State Library, Sydney, and Copy I/P23 was resold at Bonham's, New York on December 12, 2020. Bonham's also sold a partly annotated copy of the 1543 edition in the sale of June 21–30, 2023, which seems not to have been previously recorded.
207. Stephen Joffe and Veronica Buchanan, "The Vesalius Epitome of *De Humani Corporis Fabrica* of 1543. A Worldwide Census with New Findings," *Medical Research Archives* 2 (2015), 1–1, to which add the previously unknown copy sold by SophiA Rare Books of San Francisco in February 2024. Soulier, *Vésale anatomisé*, lists 76 copies of the 1564 *Examen* and 31 of the 1609 edition.
208. Zoltan I. Pozeg and Eugene S. Flamm, "Vesalius and the 1543 Epitome of His '*De humani corporis Fabrica librorum*': A

Uniquely Illuminated Copy." *Papers of the Bibliographical Society of America* 103 (2009): 199–220; Margócsy et al., *The Fabrica*, 457. Coloured copies could be ordered by individual purchasers from the printer, Kusukawa, *The Book of Nature*, 2012: 59–61.

209. Margócsy et al., *The Fabrica*, 211–12.
210. Margócsy et al., *The Fabrica*, 343; Emil Arbenz, ed. *Die Vadianische Briefsammlung der Stadtbibliothek St. Gallen, II, Mitteilungen zur vaterländische Geschichte* 25, no. 2 (1891), 247.
211. Margócsy et al., *The Fabrica*, 160, 148–50, with Jacqueline Vons, "Un exemplaire du *De humani corporis fabrica* (1543) annoté par Jan Wauters van Vieringen (Ioan. Valterius Viringus), lecteur et traducteur d'Andre Vésale," *Vesalius* 23 (2015), 38–42. The name of Vesalius, written and then deleted by the owner, seems more likely to me to be a mistake than an indication that it was owned by Vesalius.
212. Margócsy et al., *The Fabrica*, 201, 237–38, 406–7.
213. Ibid., 164, although it was soon passed onto the university library in Jena; 202–3, acquired by the Queen before 1547 and given in 1586 to her personal physician; 193–94. In the next century, copies were owned by the famous painters Velazquez and Rubens, ibid., 33; for Rubens, see also Michael W. Kwakkelstein, *Study Heads and Anatomical Studies, I. Corpus Rubenianum Ludwig Burchard, Part* X. (London and Turnhout: Harvey Miller Publishers, 2021), 17–19, 105–6.
214. Ibid., 7; Anthony Hobson, *Renaissance Book Collecting: Jean Grolier and Diego Hurtado de Mendoza, Their Books and* Bindings (Cambridge: Cambridge University Press, 1999), 81.
215. Margócsy et al., *The Fabrica*, 221, although if the date in Gwalther's hand, 1543, indicates when he acquired it, Bullinger may not have kept it for long; 342.
216. Ibid., 242–45.
217. Ibid., 158, 160, 226, 357, 379–81; the Elector of Saxony, 164; for Wittenberg bindings, 18–20.
218. Ibid., 482–86 (reading "eruditiorem", line 7). The page is copied from another publication or a fair copy of the lecture. For the copy owned by Nathan Chytraeus (1543–98), ibid., 309.
219. Ibid., 197–98 (although when it entered the college library at Puebla is unknown); Skaarup, *Early Modern Spain*, 203–23. A copy was brought by the Alcalà-educated Juan de la Fuente

to New Spain in 1561. For the critical Rodrigo de Molina, see below, and for dissection in the New World, José Pardo Tomás, "Opening Bodies in the New World. Anatomical Practices in Sixteenth-Century New Spain," in *Anatome: sezione, scomposizione, raffigurazione del corpo nell'età moderna*, edited by Giuseppe Olmi and Claudia Pancini (Bologna: Bononia University Press, 2012), 185–202.

220. Margócsy et al., *The Fabrica*, 28–30.
221. Ibid., 248.
222. Ibid., 360–61.
223. Ibid., 262–64, 459.
224. Ibid., 362–64.
225. Ibid., 203–5. Where and when Molina was teaching is unclear. He is not mentioned by Skaarup, *Early Modern Spain*; his *Institución chirurgica* was published in 1557.
226. Margócsy, "From Vesalius through Ivins."
227. John B. Oldham, "Shrewsbury School Library," *The Library, ser.* 6, no. 14 (1959), 82.
228. Margócsy et al., *The Fabrica*, 38–46; for prices, 12–16.
229. Roth, *Andreas Vesalius*; his career is summarised by Albrecht Burckhardt, *Geschichte der medizinischen Fakultät*, 463.
230. Cushing, *Bio-Bibliography*; for his career, see Michael Bliss, *Harvey Cushing, a Life in Science* (Oxford: Oxford University Press, 2005).
231. O'Malley, *Vesalius*; for his career, see the obituary by Leonard G. Wilson, "Charles Donald O'Malley 1907–1970," *Journal of the History of Medicine and Allied Sciences* 26, no. 10 (1971), 82–86, and for his work on Vesalius, Jerome J. Bylebyl, "C. D. O'Malley's Vesalius: An Appreciation", ibid., 87–100.
232. Ibid., 48–50: see also Michael Bliss, *Osler: A Life in Medicine* (Toronto: University of Toronto Press, 1999), 293–96, 363–64; Hélène Cazes, "Vesalius, 'For What He Represents': Vesaliana in William Osler's Collections," in *Andreas Vesalius and the "Fabrica" in the Age of Printing: Art, Anatomy, and Printing in the Italian Renaissance*, edited by Rinaldo Canalis and Massimo Ciavolella (Turnhout: Brepols, 2018), 259–91.
233. John Symons, *Wellcome Institute for the History of Medicine, a Short History* (1993); Robert Rhodes James, *Henry* Wellcome (London: Hodder and Stoughton, 1994).

234. Omer Steeno and Maurits Biesbrouck, "Stolen and Lost Copies of Vesalius's *Fabrica*," *Acta medico-historica adriatica* 102, no. 2 (2012), 213–36; the Russian State Library copy, now in its Museum of the Book, was removed after the last war from the Library of the Academia Leopoldina in Halle, Germany.
235. For the importance of the Lambert reprinting of the images from the original blocks in this transition, Margócsy, "Vesalius through Ivins."
236. For the former copy, Margócsy et al., *The Fabrica*, 457; for the latter, which was sold to the University of Leuven for $2,228,000, see Christie's, *De humani corporis fabrica. Vesalius' own working copy*, Sale Catalogue, online sale, 17 January to 2 February, 2024.
237. Margócsy et al., *The Fabrica*, 16–17.

CHAPTER 5

Conclusion

Few books have had such a lasting impact as the *De humani corporis fabrica* of Andreas Vesalius. In one fell swoop, it set new standards in anatomy and in the printing of scientific and illustrated books. If its strikingly beautiful images of the body were what has principally attracted the attention of scholars and connoisseurs, its message that the human body should be investigated principally through human corpses has been the motto of anatomists ever since. As Boerhaave and Albinus described in their preface to their edition three hundred years ago, Vesalius and his book take their place in the pantheon of revolts against traditional authority, whether of Galen, the Middle Ages or religion that have led up to modern science and medicine.[1]

But this iconic status has also brought with it problems, some of which are still with us today, The sheer size of Vesalius' oeuvre, whether judged by the number or the variety of his writings, to say nothing of the hundreds of large pages in Latin and the different images that he created in the *Fabrica*, has placed many obstacles in the way of understanding what he was trying to achieve. Very few doctors or surgeons today have the fluency in Latin to cope with the orotundity of his style, and although translation has made most of his work available to the non-Latinate, this has, in turn, marginalised those books that have not been turned into English or French and deprive the *Fabrica* itself of much-needed context.

V. Nutton, *Andreas Vesalius and his Fabrica, 1537–1564*, Palgrave Studies in Medieval and Early Modern Medicine, https://doi.org/10.1007/978-3-031-69565-0_5

Its technical anatomical details have deterred those without medical, and particularly anatomical, expertise from involving themselves too closely in the details of what he wrote, and only a relatively few scholars today have a broad grasp of the cultural context in which Vesalius lived and worked. There is today a greater co-operation between the two groups than in the past, but there is still much to be done. O'Malley's biography needs considerable updating and correction in the light of new discoveries, and a clearer assessment by an anatomist of what Vesalius was doing in terms that the non-anatomist can understand would in turn help the historian.

Nevertheless, the last twenty years have seen much progress. The growing availability of translations has meant that those interested in Vesalius must no longer confine themselves to gazing at the images in the *Fabrica* as icon-worshippers but can engage critically with his arguments. The increasing availability of much contemporary material put online by libraries around the world has made the task of setting him in context much easier than it was even fifteen years ago. But much still remains to be done, and, as with the editions of the *Paraphrase* discussed above, pp. 18–25, any detailed investigation is bound to reveal something new, whether about Padua, Basle or his life at court. As has been often stressed, Vesalius' rhetoric and his ability with the knife, allied to some deliberate silences, have created a barrier to a proper appreciation of his achievements by obscuring or deliberately hiding what others were doing or even what the typical activity of a professor of surgery and anatomy was. Lind's valuable collection of translations of 'pre-Vesalian' texts, published half a century ago, can now be supplemented by newly accessible material, while investigations into lecture notes and medical letters provide essential corrections to a view based on a few printed sources.[2]

But that is for the future. By contrast, this book has a more limited aim: to follow the development of Vesalius' greatest book from its beginnings in Paris, where the young Vesalius acted as assistant to Guinter von Andernach, through his first connections with Basle in 1537 to his short period of activity in Padua that effectively ended with the publication of the *Fabrica* and *Epitome* in 1543 and his decision to join the imperial court. From then on, he had little chance to dissect, although he continued to perform autopsies and act as a surgeon and physician, but that never stopped him from thinking about the results of his dissections and about how they might best be conveyed to a wider audience, something ended only by his unexpected death in 1564. His book's significance was quickly recognised. It has retained its importance ever since as marking a break with all earlier views of the anatomy of the human body, even if it was relatively quickly superseded as an up to date and accurate account of the structures of the human body. To term it an icon is appropriate, for in the twenty-first century, it has become a prized possession of libraries and similar institutions, something to be admired lying open in a showcase and consulted largely by experts.

Whether Vesalius and his book deserve this status is open to dispute. Vesalius, as many contemporaries noted, deliberately hid his debts to others and their accusations of impiety towards those from whom he had learned much struck home. Right from the start, he was aware that some would call him ungrateful for criticising Galen and Aristotle and would think little of his protestations of fidelity towards them. More damaging, because expressed more calmly, were the accusations of Falloppia and others that he too had mainly relied on dissected animals and thus often fell into the same trap as Galen. It is also true, as Michael Stolberg has pointed out, that very few major anatomical discoveries can be credited to him by comparison with Berengario or Falloppia, and even the derided Sylvius seems to have anticipated him in observing the valves in the veins.[3] But that is not surprising, for between 1534 and 1564, and particularly between 1544 and 1564, the study of anatomy developed across Europe at such enormous speed and in so many different ways. Vesalius' decision to become an imperial physician entailed abandoning almost entirely his research into the structures of the human body. It was not a foolish choice, nor, until perhaps the 1560s, is there any hint that he regretted it, but its consequence was a reduction in his opportunities to return to dissection. Vesalius was increasingly left behind by others who were either finding new structures or, like Falloppia, Coiter and Fabricius, using a more careful and more sophisticated methodology in investigating both human and animal anatomy.[4]

But even if his contributions to the understanding of the human body are not as extensive as his most ardent admirers claim, the story of the *Fabrica* provides many reasons why Vesalius deserves our attention and respect. At the very least, it was his skill in presenting his results, whether by his teaching at Padua, Bologna and Pisa or through an adroit use of the technology of printing, as Matthew Landers has noted, that helped to establish anatomy as a fundamental part of medicine, and Vesalius as, in the words of Girolamo Cardano, "the foremost exponent of the science of anatomy".[5] It is true that not all of his innovations were immediately followed. Those who argued that the occasional dissection of a human corpse and Vesalian descriptive anatomy were unhelpful to most medical practitioners were not foolish but were raising questions about the role of anatomy within the teaching of physicians that are still with us today. Even at Padua, his example of being both dissector and expositor was quickly dropped in favour of a division of labour that gave the professor the task of explaining what could be seen and putting it into the context of the living human body, a pre-Vesalian procedure that lasted into the 1560 s, if not beyond. Similarly, the sheer size and cost of the *Fabrica* may have long deterred others from producing illustrated volumes on the scale that he and Oporinus achieved. Instead, his illustrations served instead as the basis for others' shorter guides and even the humble fugitive sheets or, as with Fabricius, led them to publish a series of briefer-linked works to show different systems of the body. Even if their efforts did not always live up to Vesalius' own standards, Thomas Geminus and Christopher Langton in England, Valverde and his printer Plantin in Flanders, and Melanchthon and his acolytes in Wittenberg preached their own versions of the Vesalian message to physicians, surgeons and, among Lutherans, to children at a town gymnasium.[6]

Two different examples show how by the end of the sixteenth century anatomy had become an essential part of medical education with Vesalius as its figurehead. When the Lutheran University of Helmstedt was opened in 1576, its first professor of medicine, Johannes Boekel (1535–1605), immediately arranged for the building of a dissection hall on his own property, where he lectured on the anatomy of a male and a female corpse provided by the Duke of Brunswick. Soon afterwards, in 1585, he published his own student guide to anatomy, his *Anatome, seu descrip-tion partium humani corporis, Anatomy, or a Description of the Parts of the Human Body*. That much of this Vesalian exposition was taken over unac-knowledged from his teacher at Rostock and Copenhagen, Jacob Bording (1511–1560), is less significant than its further reprintings at Helm-stedt (1588) and Wittenberg (1589) as a valuable teaching text.[7] Shortly after this, in 1589, the leading academic physician of the day, Girolamo Mercuriale (1530–1606), who held chairs at Padua and Bologna, as well as treating both a Holy Roman Emperor and a Pope, had his portrait painted by the eminent female artist Lavinia Fontana (1552–1614).[8] She had a considerable reputation for her portraits of distinguished academics at theUniversity of Bologna, her native city, and Mercuriale enjoyed a Europe-wide reputation for his writings. He used his command of classical sources to discuss a wide variety of medical questions, from physical exercise, the subject of his most famous book, *De arte gymnastica,* through lactation and childhood diseases, to plague and poisons.[9] His lectures on Hippocrates and Galen, as well as his collections of case histories, circulated widely, and he amassed enormous wealth, including a palazzo with a celebrated art gallery in his home city of Forli. Lavinia Fontana chose to depict him in a typical academic pose, working in his well-stocked library with an open volume on the desk beside him. What is unexpected is that the book is not one of his own and deals with a subject with which, as far as can be known, he had nothing directly to do. It shows Vesalius' *Fabrica* open at one of the musclemen illustrations. Whether or not the sitter is Mercuriale, the placement of a copy of the *Fabrica* in the foreground shows how much by this stage Vesalius and his book had come to represent the summit of modern medical achievement, for the sitter must also have expected to benefit from the kudos of owning and consulting his great work.

Concentrating on the development of *Fabrica* throughout Vesalius' life, from his student days to within weeks of his death, has also allowed a closer acquaintance with the man himself that helps to understand the impact that he had on those who knew him. The constant correction and rewriting of his books, from his first publication until his last surviving annotations almost thirty years later, shows a man obsessed with getting things right. Even when deprived of the opportunity to carry out further dissections, he still wrestled with the difficulty of expressing in words the nature of the bones he had earlier seen or touched. The fruit of his early education, his Latin skill, gave him a broad vocabulary from which to choose and which he exploited to the full. It was complemented by considerable learning, especially in the Greek and Latin classics, which made him an appropriate dinner guest for an English Bishop.[10] His family connections would also have ensured that he knew what he was doing when he abandoned academia for service at court, and, at least until his final years, there is no suggestion that he regretted the move.

The later annotations furthermore confirm that he had an acute eye that could see even the tiniest flaw in a worn or badly cut block. Even as a student acting as a dissector for Guinther he could point out little things that would have been frequently invisible to someone not carrying out the cutting himself. Other anatomists, like Berengario, had anticipated some of his findings, and later others, especially Falloppia, saw things that he had missed, but this does not detract from what he had managed to see, or the importance of his injunctions to his audience to dissect and look for themselves. His insistence on seeing for oneself is accompanied by an aesthetic sense that is exemplified in the concern he had with the appearance of his books, from the size of the page and the elegance of its typography to the choice of images, whether they depicted parts of the body or surrounded an initial letter. How much Oporinus himself contributed to the enterprise remains unclear, but Vesalius was fortunate in finding a similarly ambitious collaborator who could satisfy his standards and cope with his importunities. Their friendship mattered to both men.

Vesalius was also a canny businessman. Robert Winter, Jan van Calcar, Oporinus and possibly Salviati were all engaged by Vesalius at a relatively early stage in their careers, and the first three at least made or enhanced their reputation by working for him. He was quick to see the new possibilities of printing for making his books available in revised editions and in a variety of forms to suit different markets, whether of collectors or academics, students, or the non-Latinate Germans. Vesalian anatomy, as the famous title page shows, was not something solely for doctors and surgeons.

All this took place within a very short period of time and in a remarkable outburst of energy. Niccolò Massa might grumpily complain that Vesalius had only been able to beat other like-minded anatomists into print with the *Fabrica* because of his relative freedom from other ties, but that also downplays all the other projects that Vesalius was working on during the brief period from early 1537 to summer 1543—the revised *Paraphrase, the Six Anatomical Plates,* the revision(s) of Guinter's textbook, the *Venesection letter*, and the revision of the Latin anatomical tracts of Galen, not to mention what, if anything, he had written towards a paraphrase of the complete *Al-Mansorem* or the planned book on surgery.[11] Some of his projects contributed to the making of the *Fabrica* by giving him a greater knowledge of Galen and a greater expertise with the knife, but all required a good deal of time and effort on his part.

In assessing Vesalius' achievements in the *Fabrica*, it is also easy to be misled by its very familiarity. While its general organisation goes back *to* Galen, the amount of planning that was required before a word was written was substantial. Vesalius had no precedents for such a large volume, and his earlier *Six Anatomical Plates* contained only a small proportion of the images he used later. How to organise his material was not easy. He had to decide what to display and how the relevant part should be depicted, whether singly, with other parts or sequentially on a whole or a half page. The non-anatomical illustrations for the title page, frontispiece and historiated initials show his artistic sense in their composition, as well as his wit in his choice, or at least approval, of the themes of the initials. The complexity of the arrangement for captions and references he might later dismiss as requiring little input from his typesetters other than sheer hard work, a comment which, even if justified, says something of the extent to which Vesalius was prepared to go himself in his efforts. Merely writing such a large book in so short a time is impressive even today and was reason enough for the respect in which it was held by his contemporaries. It was a book that could not be avoided even by his detractors, whose complaints that it was heavily based on the writings of others, whether justified or not, inadvertently confirm the learning that lies at its base. Other anatomists, like Falloppia and Colombo, who argued that Vesalius had not gone far enough in distancing himself from the animal-based anatomy of Galen and Aristotle, implicitly acknowledged the validity of much of what he had done.

This book has deliberately left out many of the details of Vesalius's life, particularly his work as a physician, leaving it also to others to explain his facility with the knife. Instead, it has tried to tell the story of his great book by going back to the sources, both printed and manuscript, many of them unfamiliar or disregarded, and by introducing some new contexts. In this way, I hope to have shown why the man and his book can truly be said to have revealed a new world of anatomy.

Notes

1. Hélène Cazes, "Réédition et retour au progrès: les Œuvres d'André Vésale (Leiden, Boerhaave et Albinus, 1725), acte de naissance et de renaissance de l'anatomie," in *La fabrique de la modernité scientifique. Discours et récits du progrès sous l'Ancien Régime*, edited by François Charbonneau (Oxford: Voltaire Foundation, 2015), 9–35.
2. The website aerztebriefe.de, hosted by the university of Würzburg, lists several thousands of such letters and provides for many a summary or translation in German. For a short survey of Renaissance anatomy, see Vivian Nutton, *Renaissance* Medicine (London and New York, Routledge, 2022), 245–77.

3. Michael Stolberg, *Gabrielle Falloppia 1522/3–1562. The Life and Work of a Renaissance Anatomist* (London and New York: Routledge, 2023), 57, 61.
4. For Coiter, Volker Coiter, *Externarum et internarum principaliunm humani corporis partium tabulae, atque anatomicae exercitationes, observationesque variae, novis, diversis ac artificiossimis figuris illustratae* (Nuremberg: T. Gerlatzen, 1572), with Dominik Gross and Jan Steinmetzer, *Volcher Coiter (1534–1576) und die Konstituierung ärztlicher Autorität in der Vormoderne* (Aachen: Shaker Verlag, 2006) and Hannah Murphy, A *New Order of Medicine. The Rise of Physicians in Reformation Nuremberg* (Pittsburgh: University of Pittsburgh Press, 2019), 68–96; for Fabricius, Andrew Cunnngham, "Fabricius and the 'Aristotle Project' in Anatomical Teaching and Research in Padua," in *The Medical Renaissance of the Sixteenth Century,* edited by Andrew Wear, Roger French, and Iain M. Lonie (Cambridge: Cambridge University Press, 1985), 195–222.
5. Matthew Landers, "Early Modern Dissection as a Physical Model of Organization," in *Anatomy and the Organization of Knowledge, 1500–1800*, edited by Matthew Landers and Biran Muñoz (London: Pickering and Chatto, 2012), 12–15; Girolamo Cardano, *De vita propria*, 15, in *Opera omnia* (Lyon, J. A. Huguetan and M. A. Ravaud, 1663), vol. I, 12.
6. For Lutheran schools, above, p. 219 note 32.
7. Johannes Boekel, *Anatome vel descriptio partium humani corporis ut ea in Academia Iulia quae est Helmsteti singulis annis publice praelegi ac administrari solet.* (Helmstedt: J. Lucius, 1585); Michaela Triebs, *Die medizinische Fakultät der Universität Helmstedt (1576–1810)* (Wiesbaden: Harrassowitz Verlag, 1995), 45–46; Christoph Gysel, "Jan Boekel (1535–1605) en zijn "Anatome" (1585), een plagiaat van Bordings "Physiologia" (1591)," *Acta Belgica Historiae Medicinae* 9, no. 1 (1996), 21–27; Vivian Nutton, "*Physiologia*: From Antiquity to Jacob Bording," in *Blood, Sweat and Tears—The Changing Concepts of Physiology from Antiquity into Early Modern Europe,* edited by Manfred Horstmanshoff, Helen King and Claus Zittel (Leiden and Boston: Brill, 2012), 27–40.
8. Now in the Walters Art Gallery, Baltimore; Vera Fortunati, *Lavinia Fortuna of Bologna (1552–1614)* (Milan: Electa, 1998), 82–83

and plate 17; Caroline P. Murphy, *Lavinia Fontana, A Painter and Her Patrons in Sixteenth Century Bologna* (New Haven: Yale University Press, 2003), 70–72 and plate 72. The sitter's identity is not certain, but Fontana is known to have painted at least three portraits of him, and Murphy judges the style commensurate with her work around 1588, noting also that some of the texts in the background are more obviously related to Mercuriale's interests.

9. Alessandro Arcangeli and Vivian Nutton, editors, *Girolamo Mercuriale. Medicina e cultura nell'Europa del Cinquecento* (Florence: Leo S. Olschki editore, 2008).
10. C. Donald O'Malley, *Andreas Vesalius of Brussels (1514–1564)* (Berkeley and London: University of California Press, 1964), 226. He was also a useful contact for details of diplomatic manoeuvres, ibid., 229.
11. Niccolò Massa, *Epistolae Epistolae medicinales et philosophicae* (Venice: F. Bindoni and M. Pasini, 1550), 58r; for his abandoned projects, above, pp. 199–200.

Bibliography

Primary Sources

Achillini, Alessandro. *Anatomicae annotationes magni Alexandri Achillini Bonon.* Bologna: H. De Benedictis, 1520.

Achillini, Alessandro. *De humani corporis anatomia.* Venice: G. A. Di Sabio, 1521.

Alexander of Aphrodisias, *De febrium causis….Johannes Damascenus…..De exquisita febrium curatione.* Basle: Robert Winter, 1542.

Barland, Hubert. *Epistola ad medicinae apud Lovanienese studiosam juventutem.* In Giovanni Manardo, *Medicinales epistolae.* Strasbourg: J. Schott. (1529), sigg. a.ii r–vi r.

Barland, Hubert. *Velitatio cum Arnoldo Nootz …Deque Avicennae in plaerisque horum hallucinatione. Deinde obiter agitur de venae in pleuritide sectione. Postremo Avicennae locis aliquot aliis dormitatio ostenditur, simul Galenus ab hominis morsibus aliquot defenditur.* Antwerp: Henricus Petri of Middelburg, 1532.

Baudin, Clément. *Description et demonstration des membres interieurs de l'homme et de la femme ….selon la vraye anatomie de André Wesale….Oeuvre utile & nécessaire non seulement aux Medecins & Chirurgiens, ains aussi aux Portrayeurs & Architectes.* Lyon: C. Baudin, 1560.

Bauhin, Caspar. *De corporis humani fabrica: libri IIII. Methodo anatomica in praelectionibus publicis proposita, ad Andreae Vesalii tabulas instituta: sectionibusque publicis et privatis, comprobata.* Basle: S. Henricpetri, 1590.

V. Nutton, *Andreas Vesalius and his Fabrica, 1537–1564*, Palgrave Studies in Medieval and Early Modern Medicine, https://doi.org/10.1007/978-3-031-69565-0

Benedetti, Alessandro. *Historia corporis sive Anatomice.* Venice: B. Guerraldo, 1502.

Benedetti, Alessandro. *De re medica opus insigne apprime Medicinae candidatis omnibus utile*, Basle: Heinrich Petri, 1549.

Benedetti, Alessandro. *Alessandro Benedetti,* Historia corporis sive Anatomice. Ed. and trans. by Giovanna Ferrari. Florence: Giunti, 1998.

Benivieni, Antonio. *De abditis nonnullis et mirandis morborum et sanationum causis,* Florence: F. Giunta, 1507.

Berengario da Carpi, Jacopo. *Commentaria....super Anatomia Mundini.* Bologna: H. de Benedictis, 1521.

Berengario da Carpi, Jacopo. *Isagoge breves...in anatomiam humani corporis.* Bologna: H. de Benedictis, 1522.

Berengario da Carpi, Jacopo. *Isagoge breves...in anatomiam humani corporis a communi medicorum academia usitatam, a Carpo in almo Bononiensi Gymnasio ordinariam chirurgiae docente, ad suorum scolasticorum preces in lucem datae.* Bologna: H. de Benedictis, 1523.

Berengario da Carpi, Jacopo. *A Short Introduction to Anatomy: Isagoge Breves. Translated by L. R. Lind with anatomical notes by P.G. Roofe.* Chicago: University of Chicago Press, 1959.

Bizzarri, Pietro. *Historia della guerra fatta in Ungheria*, Lyon: n.p., 1568.

Boekel, Johannes. *Anatome vel descriptio partium humani corporis ut ea in Academia Iulia quae est Helmsteti singulis annis publice praelegi ac administrari solet.* Helmstedt: J. Lucius, 1585.

Bonaveri, Domenico. *Notomie di Titiano.* Bologna: n.p., c. 168–190.

Brunfels, Otto. *Herbarum vivae ei coneb [sic,* i.e. *eicones] ad naturae imitationem, summa cum diligentia et artificio effigiatae, unà cum effectibus earundem.* Strasbourg: J. Schott, 1530.

Caius, John. *Libri aliquot graeci.* Basle: H. Froben and N. Episcopius, 1544.

Caius, John. *De libris propriis, liber unus.* London: W. Seres, 1570.

Caius, John. *An Autobibliography by John Caius, translated by Vivian Nutton.* London and New York: Routledge, 2018.

Canani, Giovanni Baptista. *Musculorum humani corporis picturata dissectio. (Ferrara 1541?).* Facsimile edition annotated by Harvey Cushing and Edward. C. Streeter. Florence: Lier, 1925.

Cardano, Girolamo. *Libelli quinque.* Nuremberg: G. Petreius, 1547.

Cardano, Girolamo. *Opera omnia.* Lyon: J. A. Huguetan and M. A. Ravaud. 1663.

Coiter, Volker. *Externarum et internarum principaliunm humani corporis partium tabulae, atque anatomicae exercitationes, observationesque variae, novis, diversis ac artificiossimis figuris illustratae.* Nuremberg: T. Gerlatzen, 1572.

Colombo, Realdo. *De re anatomica.* Venice: N. Bevilacqua, 1559.

Colombo, Realdo. *De re anatomica libri XV; Anatomia. Edited with translation and commentary by G. Baldo.* Paris: Les Belles Lettres, 2014.

Cornarius, Janus. *Universae rei medicae ἐπιγραφή seu enumeratio, compendio tractata.* Basle: Froben, 1529.

Cornarius, Janus. *Universae rei medicae ἐπιγραφή seu enumeratio, compendio tractata, eius auctore recognita atque aliquot locis aucta.* Basle: Froben, 1534.

Crassus, Junius Paulus. *Medici antiqui graeci.* Basle: P. Perna, 1581.

Cuneo, Gabriele. *Apologia Francisci Putei pro Galeno in anatome examen.* Venice: Francesco de' Franceschi, 1564.

Da Monte, Giovanni Battista. *In primi libri Canonis Avicennae primam Fen, profundissima commentaria.* Venice: V. Valgrisi and B. Constantino, 1558.

Da Monte, Giovanni Battista. *Medicina universa.* Frankfurt: V. Marnius and J. Aubrius, heirs of A. Wechel, 1587.

De Goes, Damian. *Fides, mores, religioque Aethiopum.* Louvain: R. Rescius, 1549.

Dolet, Estienne. *Carminum libri quatuor.* Lyon: [the author], 1538.

Dryander, Johannes. *Anatomia capitis humani.* Marburg: E. Cervicornus, 1536.

Drivère, Jérôme. *De missione sanguinis in pleuritide...commentarius de victu ab arthriticis morbis vindicante, ubi quam male diris illis cruciatibus sit a neotericis hactenus provisum, ostenditur.* Louvain: Rutger Rescius, 1532.

Du Fail, Noel. *Contes facétieuses.* Paris: P. Daffis, 1874.

Erasmus, Desiderius, *Opus epistolarum Des. Erasmi Roterodami*, edited by (Oxford: Clarendon Press, 1906–1947).

Estienne, Charles. *De dissectione partium corporis humani.* Paris: S. de Colines, 1545.

Estienne, Charles. *La dissection des parties du corps humain.* Paris: S. de Colines, 1546.

Falloppia, Gabriele. *Observationes anatomicae.* Venice: Marcantonio Olmo and Gratioso Perchachino, 1561.

Fernel, Jacques. *De naturali parte medicinae.* Paris: A. Wechel, 1542.

Fernel, Jacques. *The* Physiologia *of Jean Fernel (1567). Translated and Annotated by John M. Forrester with an Introduction by John Henry and John M. Forrester.* Philadelphia: American Philosophical Society, 2003.

Fuchs, Leonhart. *Compendiaria ac succincta admodum in artem medendi....introductio.* Hagenau: J. Secerius, 1531.

Fuchs, Leonhart. *Compendiaria in artem medendi introductio, diligenter recognita, & locis multis quam antea auctior.* Strasbourg: J. Albertus, 1534.

Fuchs, Leonhart. *De historia stirpium commentarii insignes.* Basle: M. Isingrin, 1542.

Fuchs, Leonhart. *De humani corporis fabrica epitomes....pars prima.* Lyon: J. Frellon, 1551.

Fuchs, Leonhart. *De humani corporis fabrica epitome....pars altera.* Lyon: J. Frellon, 1555.

Galen. *Opera omnia.* Venice: heirs of Aldus, 1525.
Galen. *Libri anatomici.* Bologna: G. B. Phaelli, 1529.
Galen. *De anatomicis administratinibus.* Paris: S. De Colines, 1531.
Galen. *Opera omnium utilissima.* Basle: Andreas Cratander, 1536.
Galen. *Opera omnia, ad fidem complurium & perquam vetustorum exemplariorum ita emendata atque restituta, ut nunc primum nata, atque in lucem aedita, uideri possint.* Basle: Cratander, 1538.
Galen. *Opus medicum practicum.* Basle: H. Froben and N.Episcopius, 1537.
Galen. *Opera quae extant omnia.* Venice: Giunta, 1541–1542.
Galen. *Opera omnia.* Venice: G. Farri and brothers, 1541–1545.
Galen. *Opera.* Basle: H. Frobenius and N. Episcopius, 1542.
Galen. *Galen On Anatomical Procedures: Translation of the Surviving Books by Charles Singer.* London, Oxford and New York: Oxford University Press, 1956.
Galen. *Galen on Anatomical Procedures: The Later Books, translated by W. H. L. Duckworth, M. C. Lyons, and B. Towers.* Cambridge: Cambridge University Press, 1962.
Galen. *On the Usefulness of the Parts of the Human Body,* trans. M. T. May, 2 vols. Ithaca: Cornell University Press, 1968.
Gelenius, Sigismund. *Notitia utraque cum orientis tum occidentis ultra Arcadii Honoriique caesarum tempora.* Basle: Hieronymus Froben and Nicolaus Episcopius, 1552.
Geminus, Thomas. *Compendiosa totius anatomiae delineatio aere exarata.* London: J. Herford, 1545.
Geminus, Thomas. *Compendiosa totius anatomiae delineatio aere exarata.*
London: N. Hyll for T. Geminus.
Gemusaeus, Hieronymus. *In libros Pauli Aeginetae omnes, annotationes nunc et primae concinnatae.* Basle: Andreas Cratander, 1543.
Gessner, Conrad. *Bibliotheca Universalis.* Zurich: C. Froschover, 1545.
Gessner, Conrad. *Appendix Bibliothecae Universalis.* Zurich: C. Froschover, 1555.
De Goes, Damian. *Fides, Mores, Religioque Aethiopum.* Louvain: Rutger Rescius, 1540.
Giovio, Paolo. *Historiae sui temporis,* Florence: L. Torrentino, 1550.
Gryllus, Lorenz. *De sapore.* Prague: G. Melantrich, 1566.
Guinter, Johann. *De medicina veteri et nova facienda.* Basle: ex officina Henricpetrina, 1571.
Guinter, Johann. *Institutionum anatomicarum secundum Galeni sententiam ad candidatos medicinae libri quatuor.* Paris: S. de Colines, 1536.
Guinter, Johann. *Institutionum anatomicarum secundum Galeni sententiam ad candidatos medicinae libri quatuor.* Basle: Balthasar Lasius and Thomas Platter, 1536.

Guinter, Johann. *Institutionum anatomicarum secundum Galeni sententiam libri quatuor.* Basle: Robert Winter, 1539.

Guinter, Johann and Andreas Vesalius. *Institutionum anatomicarum secundum Galeni sententiam... libri ...Ab Andrea Wesalio Bruxellensi auctiores et emendatiores redditi.* Venice: D. Bernardino, 1538.

Guinter, Johann and Andreas Vesalius. *Institutionum anatomicarum secundum Galeni sententiam... libri ...Ab Andrea Wesalio Bruxellensi auctiores et emendatiores redditi.* Venice: M. Sessa, n.d. (1540).

Guinter, Johann. *Principles of Anatomy According to the Opinion of Galen,* translated by Vivian Nutton. London and New York: Routledge, 2019.

Hartmann, Alfred, Beat Jenny, Reinhard Bodenmann and Lorenz Heiligensetzer. *Die Amerbachkorrespondenz,* 11 vols in 14. Basle: Universitätsbibliothek, (1942–2010).

Henerus, Renatus. *Adversus Jacobi Sylvii depulsionum anatomicarum calumniarum, pro Andrea Vesalio apologia.* Venice: n.p., 1555.

Hippocrates. *Libri omnes, ad vetustos codices summon studio collati et restaurati.* Basle: H. Froben and N. Episcopius, 1538.

Ingrassia, Gian Filippo. *De tumoribus praeter naturam.* Naples: M. Cancer, 1553.

Koran. *Machumetis saracenorum principis eiusque successorum vitae, ac doctrina, ipseque alcoran, ... quae ... Petrus abbas Cluniacensis ... ex Arabica lingua in Latinam transferri curavit. Basle: J. Oporinus.*

Laguna, Andreas. *Anatomica methodus, seu de sectione humani corporis contemplatio.* Paris: J. Kerver, 1536.

Landi, Bassiano. *De humana historia, uel singularum hominis partium cognitione.* Basle: J. Oporinus, 1542.

Langton, Christopher. *An Introduction into Physicke, with an Universal Dyet.* London: E(dward) W(hitchurch), undated.

Langton, Christopher. *A very brefe treatise ordrely declaring the principal partes of physick, that is to say: - Thynges naturall, Thynges not naturall, Thynges against nature.* London: Edward Whitchurch, 1547.

Leoniceno, Niccolò. *De Plinii et aliorum in medicina erroribus.* Ferrara: L. de Rubeis and A. de Grassis, 1492.

Leveling, Heinrich Palmaz. *Anatomische Erklärung der Original-Figuren von Andreas Vesal: Samt einer Anwendung der Winslowischen Zergliederungslehre in sieben Büchern.* Ingolstadt: Anton Attenkhouer, 1783.

Maggi, Bartolomeo. *De vulnerum sclopetorum, et bombardarum curatione.* Bologna: B. Bonardus, 1552.

Manardo, Giovanni. *Epistolae medicinales.* Basle: J. Bebel, 1535.

Marcolini, Francesco. *Le sorti.* Venice: Francesco Marcolini, 1541.

Maschenbauer, Andreas. *Andreae Vesalii Bruxellensis, Deß Ersten, Besten Anatomici, Zergliederung Deß Menschlichen Körpers. Auf Mahlerey und Bildhauer-Kunst gericht.* Augsburg: Andreas Maschenbauer, 1706.

Massa, Niccolò. *Anatomiae liber introductorius*. Venice: F. Bindoni and M. Pasini, 1536.
Massa, Niccolò. *Epistolae medicinales et philosophicae*. Venice: F. Bindoni and M. Pasini, 1550.
Mattioli, Pietrandrea, *Epistolae medicinales*. Venice: Vincenzo Valgrisi, 1561.
Melanchthon, Philip. *Briefwechsel. Kritische und kommentierte Gesamtausgabe*. Stuttgart and Bad Cannstatt: Verlag Frommann-Holzboog, 1977.
Mondino de' Liuzzi. *Anothomia*, edited by Petro P. Giorgi and Giorgio F. Pasini. Bologna: Istituto per la Storia dell' Università di Bologna, 1992.
Morel, Théodore. *Tredecim orationes encomiasticae habitae Parisiis, Anno M.D.XXIII*. Paris: N. Savetier, 1523.
Panizza, Ludovico. *De venaesectione in inflammationibus*. Venice: F. Camozzi, 1561.
Paul of Aegina, *Libri septem*. Basle: Andreas Cratander, 1538.
Plato. *Opera omnia*. Basle: Johann Walder, 1534.
Platter, Felix. *De corporis humani structura et usu*. Basle: A. Froben, 1583.
Purkircher, Georg. *Opera quae supersunt omnia*, edited by Magnus Okál. Budapest: Akadémiai Kiadó, 1588.
Puteus, Franciscus. *Apologia in anatome pro Galeno, contra Andream Vesalium Bruxellensem*. Venice: Francesco di Portinari di Tridino, 1562.
Rabelais, François. *Hippocratis et Galeni libri aliquot ex recognitione Francisci Rabelaesi*. Lyon: S. Gryphius, 1538.
Ruini, Carlo. *Anatomia del cavallo, infermità, e suoi rimedii*. Bologna: Eredi di Giovanni Rossi, 1598.
Salviati. Giuseppe. *Regola di far perfettamente col compass la voluta et del capitello ionico et d'ogn' altra sorte*. Venice: Francesco Marcolini, 1552.
Sylvius, Jacobus. *In Hippocratis et Galeni physiologiae partem anatomicam Isagoge*. Paris: J. Hulpeau, 1555.
Sylvius, Jacobus. *Methodus sex librorum Galeni in differentiis et causis morboru, et symptomatum in tabellas sex ordine sua coniecta... de signis omnibus medicis...commentarius*. Paris: C. Wechel, 1539.
Sylvius, Jacobus. *Ordo et ordinis ratio in legendis Hippocratis et Galeni libris*. Paris: C. Wechel, 1539.
Sylvius, Jacobus. *Vaesani cujusdam calumniarum in Hippocratis Galenique anatomiam depulsio*. Paris: C. Barbé and the widow of I. Gazellus, 1551.
Tagault, Jean. *De chirurgica institutione*. Paris: Chrestien Wechel, 1543.
Theophilus Protospatharius. *De corporis humani fabrica libri quinque a Junio Paulo Crasso Patavino in Latinam orationem conversi*. Venice: O. Scotus, 1536.
Theophrastus, *Opera Omnia*. Basle: J. Oporinus, 1541.

Thorer, Alban. *Abupetri Rhazae Maomethi ....summi medici opera exquisitiora. Per Gerardum Toletanum medicum Cremonensem, Andream Vesalium Bruxellensem, Albanum Torinum Vitoduranum Latinitate donata.* Basle: Heinrich Petri, 1544.

Tortebat, François, *Abregé d'anatomie accommodé aux arts de peinture et de sculpture,* Paris: Tortebat, 1667.

Valla, Giorgio. *De expetendis et fugiendis rebus.* Venice: Aldus Manutius, 1501.

Valverde de Hamusco, Juan. *Historia de la composicion del cuerpo humano.* Rome: A. Salamanca and A. Lafreri, 1556.

Vesalius, Andreas. *Paraphrasis in nonum librum Rhazae medici Arabis clariss. ad Regem Almansorem, de singularum corporis partium affectuum curatione.* Louvain: R. Rescius, 1537.

Vesalius, Andreas. *Paraphrasis in nonum librum Rhazae medici Arabis clariss. ad Regem Almansorem, de singularum corporis partium affectuum curatione.* Basle: R. Winter, 1537?.

Vesalius, Andreas. *Tabulae anatomicae sex.* Venice: B. Vitalis, 1538.

Vesalius, Andreas. *Epistola, docens venam axillarem dextri cubiti in dolore laterali secandam.* Basle: R. Winter, 1539.

Vesalius, Andreas. *De humani corporis fabrica.* Basle: J. Oporinus, 1543.

Vesalius, Andreas. *Suorum de humani corporis fabrica, Librorum epitome.* Basle: J. Oporinus, 1543.

Vesalius, Andreas. *Von der menschen Cörpers Anatomey, ein kurtzer aber vast nützer Ausz D. Andree Vesalii... Büchern, von ihm selbs in Latein bescriben, unnd durch D. Albanum Torinum verdolmetscht.* Basle: J. Oporinus, 1543.

Vesalius, Andreas. *Epistola, rationem modumque propinandi radicis Chynae decocti....pertractans.* Basle, J. Oporinus, 1546.

Vesalius, Andreas. *Anatomia deudsch; ein kurzere Auszug der Beschreibung aller Glider menschlichs Leybs.* Nuremberg: J. P. Fabricius, 1551.

Vesalius, Andreas. *De humani corporis fabrica.* Lyon: J. de Tournes, 1552.

Vesalius, Andreas. *De humani corporis fabrica.* Basle: J. Oporinus, 1555.

Vesalius, Andreas. *Anatomicarum Gabrielis Falloppii observationum examen.* Venice: Francesco de' Franceschi, 1564.

Vesalius, Andreas. *De humani corporis fabrica.* Venice: Francesco de' Franceschi, 1568.

Vesalius, Andreas. *Anatomia.* Venice: Giovanni Antonio and Giacomo de' Franceschi, 1604,

Vesalius, Andreas. *Opera omnia Anatomica et Chirurgica.* Leiden: J. de Vivie and J. and H. Verbeek, 1725.

Vesalius, Andreas. *André Vésale, La Structure du corps humain*, translated by Jules- Émile Verschaffelt. Haarlem: Enschede & Sons, 1924.

Vesalius, Andreas. *Andreas Vesalii Bruxellensis "Icones anatomica"*. Munich and New York: Library of the University of Munich and the New York Academy of Medicine, 1935.

Vesalius, Andreas. *De humani corporis fabrica,* translated by Vasiliy N. Ternovsky and Sergei P. Shestakov, Moscow: USSR Academy of Sciences, 1950–54.

Vesalius, Andreas. *Vesalius on the Human Brain,* translated by Charles Singer. London: Oxford University Press for the Wellcome Historical Medical Museum, 1952.

Vesalius, Andreas. *The* Epitome *of Andreas Vesalius,* translated by Levi R. Lind. New York: MacmIllan Co., 1959.

Vesalius, Andreas. *Vesalius, On the Fabric of the Human Body,* translated by William F. Richardson and John B. Carman. Novato, CA: Norman Publishing, 1998–2009.

Vesalius. Andreas. *Andreas Vesalius, The Fabric of the Human Body*, translated by Daniel H. Garrison and Malcolm H. Hast. Basle: Karger, 2014.

Vesalius, Andreas. *Vesalius:* The China Root Epistle. *A New Translation and Critical Edition by Daniel H. Garrison.* Cambridge: Cambridge University Press, 2015.

Vesalius, Andreas. *Examen des* Observations anatomiques *de Gabriel Fallope. Édition de David Soulier.* Paris : Classiques Garnier, 2023.

Zerbi, Gabriele. *Liber anathomiae corporis humani et singulorum membrorum illius.* Venice: B. Locatello, 1502.

Secondary Literature

Alarcon, Carolina. "Imitation is the Sincerest Form of Innovation: Valverde Reconsidered." *The Sixteenth Century Journal* 53(1), 2022: 1–39.

Andretta, Elisa. *Roma medica. Anatomie d'un système médical au XVIe siècle.* Rome: École française de Rome, 2011.

Anson, Barry J. "The Initial Letters of the 1555 Edition of Vesalius's *Fabrica.*" *Quarterly Bulletin of the Northwestern University Medical School* 19, 1943: 326–335.

Antonioli, Roland. *Rabelais et la médecine.* Geneva: Droz, 1976.

Arbenz, Emil, ed. *Die Vadianische Briefsammlung der Stadtbibliothek St. Gallen, II, Mitteilungen zur vaterländische Geschichte* 25(2), 1891.

Arcangeli, Alessandro and Vivian Nutton, eds. *Girolamo Mercuriale. Medici na e cultura nell'Europa del Cinquecento.* Florence: Leo S. Olschki editore, 2008.

Baader, Gerhard. "Jacques Dubois as a Practitioner." In *The Medical Renaissance of the Sixteenth Century,* edited by Andrew Wear, Roger K. French and Iain M. Lonie, 146–154. Cambridge: Cambridge University Press, 1985.

Baker, Tawrin. "Why All this Jelly? Jacopo Zabarella and Hieronymus Fabricius ab Aquapendente on the Usefulness of the Vitreous Humour." In

Early Modern Medicine and Natural Philosophy, edited by Peter Distelzweig, Benjamin Goldberg and Evan R. Ragland, 59–88. Wiesbaden: Springer, 2016.

Barcia Goyanes, Juan José. *El mito de Vesalio*. Valencia: Universidad de Valencia, 1994.

Bernabò, Massimo. "La Rinascenza Macedone e l'arte dei bendaggi." *Kos* I, (5), 1984: 45–47.

Biesbrouck, Maurits. "Rutgerus Giffeus, un nouvel ami d'André Vésale, qui fait de la Paraphrasis de la BIUM un exemplaire unique au monde." *Histoire des Sciences médicales* 45(2), 2011: 196–199.

Biesbrouck, Maurits and Omer Steeno. "Andreas Vesalius (1514–1564) and Christophorus Pfluegel (1520–after 1587) and the Role of the Students in the Siege of Leuven in 1542." In *In the Shadow of Vesalius*, edited by Rob Van Hee, 24–35. Antwerp and Apeldoorn: Garant, 2020.

Biesbrouck, Maurits and Omer Steeno (on-line). "The *Paraphrasis ad librum nonum librum Rhazae*, first book by Vesalius." In *La Fabrique de Vésale et autres textes,* on line at www,biusanté.parisdescartes.fr/vesale/pdf/para/pdf.

Maurits Biesbrouck, Theodoor Goddeeris and Omer Steeno "Andreas Vesalius in Bologna." In *Tijdgenooten uit de leefwereld van Andreas Vesalius*, edited by Rob Van Hee and Maurits Biesbrouck, 123–140. Antwerp and Apeldoorn: Garant, 2018.

Biffis, Mattia. "Porta, Giuseppe." In *Dizionario biografico degli italiani* (1960–). Rome: Treccani 85, 2001: 98–102.

Bigotti, Fabrizio. *Physiology of the Soul. Mind, Body and Matter in the Galenic Tradition of the Late Renaissance (1550–1630)*. Turnhout: Brepols, 2020.

Bigotti, Fabrizio. "Commenting on Aristotle with a Knife. The Heretical Anatomies of Bassiano Landi." *NTM, n.s.*31 2023: 1–26.

Biow, Douglas. *Doctors, Ambassadors, and Secretaries: Humanism and Professions in Renaissance Italy*. Chicago and London: The University of Chicago Press, 2002.

Bliss, Michael. *Osler: A Life in Medicine*. Toronto: University of Toronto Press, 1999.

Bliss Michael. *Harvey Cushing, a Life in Science*. Oxford: Oxford University Press, 2005.

Bolgar, T. Robert. "Rabelais' Edition of the *Aphorisms* of Hippocrates." *Modern Language Review* 35(1), 1940): 62–66.

Bols, Peter E. J. "How Andreas Vesalius Inspired Early Veterinary Authors." in *In the Shadow of Vesalius,* edited by Rob Van Hee, 249–265. Antwerp and Apeldoorn: Garant, 2020.

Boudon-Millot, Véronique. "Vésale lecteur de Galien et d'Hippocrate." In *La* Fabrique *de Vésale. La mémoire d'un livre,* edited by Jacqueline Vons, 11–26. Paris: Bibliothèque interuniversitaire de Santé, 2016.

Bouras-Vallianatos, Petros. "Galen in Byzantine Medical Literature." In *Brill's Companion to the Reception of Galen*, edited by Petros Bouras-Vallianatos and Barbara Zipser, 86–110. Leiden and New York: Brill, 2019.

Branca, Vittorio (ed.). *Giorgio Valla tra scienza e sapienza.* Florence: Leo S. Olschki, 1981.

Bresadola, Marco. "*Giovanni Battista Canani : l'umanesimo medico e la riforma dell'anatomia rinascimentale.*" *Schifanoia* 50/51, 2016: 13–20.

Brinkhus, Gerd and Claudine Pachnicke (eds.). *Leonhart Fuchs 1501–1566. Mediziner und Botaniker.* Tübingen: Stadtmuseum, 2001.

Brockliss, Laurence and Colin Jones. *The Medical World of Early Modern France.* Oxford: Clarendon Press, 1997.

Bubb, Claire. *Dissection in Classical Antiquity: A Social and Medical History,* Cambridge: Cambridge University Press, 2022.

Burckhardt, Albrecht. *Geschichte der medizinischen Fakultät zu Basel 1460–1900.* Basle: Friedrich Reinhard, 1917.

Burckhardt, Andreas. *Johannes Basilius Herold. Kaiser und Reich im protestantischen Schrifttum des Basler Buchdrucks um die Mitte des 16. Jahrhunderts.* Basle and Stuttgart: Helbing & Lichtenhahn, 1967.

Bylebyl, Jerome J. "*C. D. O'Malley's Vesalius: An Appreciation*", *Journal of the History of Medicine and Allied Sciences* 26, 1971: 87–100.

Bylebyl, Jerome J. "Interpreting the *Fasciculo* Anatomy Scene." *Journal of the History of Medicine and Allied Sciences* 45, 1990: 285–316.

Caiati, Martia R., Maria Cecilia Fabbri, Robrecht Van Hee, Beatrice Campanella, Stefano Legnaioli, Stefano Pagnotta Stefano, Vincenzo Palleschi and Francesco Poggialini. "Vesalius' "Philosopher", a Recently Found Drawing by Jan Steven van Calcar." In *In the Shadow of Vesalius*, edited by Rob Van Hee, 13–24. Antwerp and Apeldoorn: Garant, 2020.

Canalis, Rinaldo F. "Gabrielle Falloppia; Vesalius's Admirer and First Critic." In *Andreas Vesalius and the "Fabrica" in the Age of Printing: Art, Anatomy, and Printing in the Italian Renaissance,* edited by Rinaldo F. Canalis and Massimo Ciavolella, 171–200. Turnhout: Brepols, 2018.

Canalis, Rinaldo F. (2018a) "Vesalius' Methods in the Production of the *Fabrica* with Emphasis on the Neuroanatomy Chapters." In *Andreas Vesalius and the "Fabrica" in the Age of Printing: Art, Anatomy, and Printing in the Italian Renaissance*, edited by Rinaldo Canalis and Massimo Ciavolella, 131–170. Turnhout: Brepols, 2018.

Carlino, Andrea. *Books of the Body: Anatomical Ritual and Renaissance Learning.* Chicago: Chicago University Press, 1999.

Carlino, Andrea. "Medical Humanism: Rhetoric and Anatomy at Padua, circa 1540." In *Rhetoric and Medicine in Early Modern Europe,* edited by Stephen Pender and Nancy S. Struever, 111–128. Farnham and Burlington, VT: Ashgate, 2012.

Carlino, Andrea. *Paper Bodies. A Catalogue of Anatomical Fugitive Sheets 1538–1687, Medical History Supplement 19.* London: The Wellcome Institute, 1999.

Castiglioni, Arturo. "The Attack of Franciscus Puteus on Andreas Vesalius and the defence by Gabriel Cuneus." *Yale Journal of Biology and Medicine* 16, 1943: 135–142.

Catani, Marco and Stefano Sandrone. *Brain Renaissance from Vesalius to Modern Neuroscience.* Oxford: Oxford University Press, 2015.

Cavanagh, G. S. Terence. "A New View of the Vesalian Landscape." *Medical History* 27, 1983: 77–79.

Cazes, Hélène. "Naufrages et hommages sur l'Île d'André Vésale: reconnaissances d'une autorité contre la tradition." In *Towards the Authority of Vesalius: Studies on Medicine and the Human Body from Antiquity to the Renaissance and Beyond*, edited by Erika Gielen and Michèle Goyens, 437–467. Turnhout: Brepols, 2018.

Cazes, Hélène. "Réédition et retour au progrès : les Œuvres d'André Vésale (Leiden, Boerhaave et Albinus, 1725), acte de naissance et de renaissance de l'anatomie." In *La fabrique de la modernité scientifique. Discours et récits du progrès sous l'Ancien Régime*, edited by François Charbonneau, 9–35. Oxford: Voltaire Foundation, 2015.

Cazes, Hélène. "Vesalius, 'for What He Represents': Vesaliana in William Osler's Collections." In *Andreas Vesalius and the "Fabrica" in the Age of Printing: Art, Anatomy, and Printing in the Italian Renaissance*, edited by Rinaldo Canalis and Massimo Ciavolella, 259–291. Turnhout: Brepols, 2018.

Hélène Cazes and Carlino, Andrea. "Plaisir de l'anatomie, plaisir du livre: *La dissection des parties du corps humain* par Charles Estienne (Paris, 1546). *Cahiers de l'Association Internationale des Études Françaises* 55, 2003: 251–269.

Cazort, Mimi, Monique Kornell, and Kenneth B. Roberts. *The Ingenious Machine of Nature.* Ottawa: National Gallery of Canada, 1996.

Cellamare, David. (2018), "History of a Productive Failure. The Use of Andreas Vesalius's *De humani corporis fabrica* in some Sixteenth-century Books on the Soul." In *Towards the Authority of Vesalius: Studies on Medicine and the Human Body from Antiquity to the Renaissance and Beyond*, edited by Erika Gielen and Michèle Goyens, 403–436. Turnhout: Brepols, 2018.

Chambers, David S. "A Mantuan in London in 1557: Further Research on Annibale Litolfi." In *England and the Continental Renaissance. Essays in Honour of J. B. Trapp*, edited by Edward Chaney and Peter Mack, 73–108. Woodbridge: The Boydell Press, 1990.

Charreaux, Stéphanie and Jérôme Van Wijland. "Recensement et déscription des exemplaires de la première édition du De Fabrica (1543) conservés en France dans les bibliothèques publiques." In *La* Fabrique *de Vesale. La mémoire d'un livre*, edited by Jacqueline Vons, 253–312. Paris : Bibliothèque interuniversitaire de Santé, 2016.

Choné, Paulette. "Le corps pathétique de Gabriel Salmon (v. 1485–1542." In *La Représentation du corps à la Renaissance,* péristyles 42, edited by Paul Vert, Catherine Charoy and Mireille Mazet, 105–116. Nancy: Musée des Beaux Arts, 2013.

Choulant, Ludwig. *History and Bibliography of Anatomic Illustration, translated and annotated by Mortimer Frank.* Chicago: University of Chicago, 1920.

Chrisman, Miriam Usher. *Lay Culture, Learned Culture. Books and Social Change in Strasbourg, 1480–1599*, New Haven and London: Yale University Press, 1982.

Christie's, *De humani corporis fabrica. Vesalius' own Working Copy,* Sale Catalogue, online sale, 17 January to 2 February, 2024.

Clark, Harry. "Foiling the Pirates.The Preparation and Publication of Andreas Vesalius's De humani corporis fabric." Library Quarterly 51, 1981: 301–311.

Clayton, Martin and Philo, Ron. *Leonardo da Vinci: Anatomist.* London: Royal Collection Enterprises Ltd, 2012.

Cockx-Indestege, Elly. *Andreas Vesalius, A Belgian Census: Contribution towards a New Edition of H.W. Cushing's Bibliography. Brussels: Royal Library*, 1994.

Cockx-Indestege, Elly. "Jaspar Laet Senior, Govaert Bac en Pierre le Caron. Een pronosticatie voor het Jaar 1500." In *Hellinga, Festschrift, Festbundel, Mélanges*: *Forty-three Studies in Bibliography presented to Prof. Dr. Wytze Hellinga,* edited by Arnold R.A. Croiset van Uchelen, 74–80. Amsterdam: Nico Israel, 1980.

Collins, Minta. *Medieval Herbals. The Illustrative Tradition.* London: The British Library and The University of Toronto Press, 2000.

Compier, Abdul Haq. "Rhazes in the Renaissance of Andreas Vesalius." *Medical History* 56, 2012: 3–25.

Concasty, Marie-Louise. *Commentaires de la Faculté de médecine de l'Université de Paris (1516–1560).* Paris: Imprimerie Nationale, 1964.

Cunningham, Andrew. "Fabricius and the 'Aristotle Project' in Anatomical Teaching and Research in Padua." In *The Medical Renaissance of the Sixteenth Century,* edited byAndrew Wear, Roger French, and Iain M. Lonie, 195–222. Cambridge: Cambridge University Press, 1985.

Cunningham, Andrew and Tamara Hug. *Focus on the Frontispiece of the* Fabrica *of Vesalius*, Cambridge: Wellcome Unit for the History of Medicine, 1994.

Cushing, Harvey W. *A Bio-Bibliography of Andreas Vesalius,* Hamden, CT, and London: Archon Books, 1962.

De Ferrari, Antonio. "Corti, Matteo." In *Dizionario biografico degli italiani* (1960–). Rome: Treccani 29, 1983: 795–797.

De Vocht, Henry. *History of the Foundation and the Rise of the Collegium Trilingue Lovaniense 1517–1550.* Louvain: Bibliothèque de l'Université, 1951–1955.

Dirix, Theo. *In Search of Andreas Vesalius. The Quest for the Lost Grave.* Tielt: Lannoo Campus, 2014.

Donaldson, Iain M. L. "Andreae Vesalii Bruxellensis Icones anatomicae. Part 2." *The Journal of the Royal College of Physicians of Edinburgh* 42(3): 280–282.

Donaldson, Iain M. L. "Two States of Some Plates in the *Compendiosa* of Thomas Geminus (1545)." *The Library* 11(1), 2010: 89–104.

Drizenko, Antoine. Les "*Institutions Anatomiques* de Jean Guinter d'Andernach (1487–1564) et André Vésale (1524–1564)." *Histoire des sciences médicales* 45, 2011: 321–328.

Ebgi, Raphael. "Cum barbaris decertare. *Sull'umanesimo medico di Niccolò Leoniceno e sulla sua controversia col Poliziano.*" *Schifanoia* 50/51, 2016: 55–62.

Eisenstein, Elizabeth. *The Printing Press as an Agent of Change.* Cambridge: Cambridge University Press, 1979.

Elaut, Léon. "De derde en de vierde druk van Vesalius' Fabrica: 1568, 1604." *Scientiarum Historia* 6, 1974: 124–34.

Eriksson, Ruben. *Andreas Vesalius' First Public Anatomy at Bologna 1540. An Eye-witness Report by Baldasar Heseler medicinae scolaris together with his Notes on Matthaeus Curtius' Lectures on Anatomia Mundini.* Uppsala and Stockholm: Almqvist and Wiksell, 1959.

Evans, Richard J. W. *Rudolf II and his World. A Study in Intellectual History 1576–1612,* Oxford: Clarendon Press, 1984.

Facciolati, Jacopo. *Fasti gymnasii Patavini ab anno mdxvii quo restitutae scholae sunt ad mdcclvi.* Padua: Giovanni Manfrè, 1757.

Faller, Adolf. *As iniciais da II ª ediçião (1555) de obra 'De humani corporis fabrica Lib;ri septem' da Andrea Vesalio,* Rio de Janeiro: CIBA, 1951.

Febvre, Lucien and Henri-Jean Martin. *L'apparition du livre.* Paris: Albin Michel, 1999.

Ferrari, Giovanna (ed.). *Alessandro Benedetti,* Historia corporis sive Anatomice. Florence: Giunti, 1998.

Ferrari, Giovanna. *L'esperienza del passato. Alessandro Benedetti, filologo e medico umanista.* Florence: L. S. Olschki, 1996.

Ferrari, Giovanna. "Public Anatomy Lessons and the Carnival; the Anatomy Theatre of Bologna." *Past and Present* 117, 1987: 50–106.

Ferrari, Giovanna. "Tra medicina e chirurgia: la rinascita dell'anatomia e la dissezione come *spettacolo.*" In *Il Rinascimento italiano e l'Europa: Volume quinto, Le Scienze.* edited by Antonio Clericuzio and Germana Ernst, 41–68. Rome: Fondazione Cassamarca, 2008.

Ferretto, Silvia. *Maestri per il metodo di trattar le cose. Bassiano Lando, Giovan Battista da Monte e la scienza della medicina nel xvi secolo.* Padua: CLEUP, 2012.

Findlen, Paula. "Anatomy Theatres, Botanic Gardens, and Natural History Collections." In *The Cambridge History of Science, vol. 3, Early Modern Science,* edited by Katharine Park and Lorraine Daston. 272–289. Cambridge: Cambridge University Press, 2006.

Findlen, Paula. "The Formation of a Scientific Community: Natural History in Sixteenth-century Italy." In *Natural Particulars: Nature and the Disciplines in Renaissance Europe,* edited by Anthony Grafton and Nancy G. Siraisi, 369–400. Cambridge, MA: The MIT Press, 1999.

Fontaine, Marie-Madeleine. *Libertés et savoirs du corps à la Renaissance,* Caen : Paradigme, 1993.

Fortuna, Stefania. "L'insegnamento di Nicolò Leoniceno e la sua incidenza nella storia della medicina italiana ed europea." In *Nicolò Leoniceno 1428–1524 Un umanista veneto nella storia della Medicina. Atti del Convegno in Lonigo per il 590 anniversario della nascita*, edited by Alberto Lonigo, 81–96. Contro: Riccardo editore, 2019.

Fortunati, Vera. *Lavinia Fortuna of Bologna (1552–1614).* Milan: Electa, 1998.

French, Roger K. "Berengario da Carpi and the Use of Commentary in Anatomical Teaching." In *The Medical Renaissance of the Sixteenth Century,* edited by Andrew Wear, Roger French and Iain M. Lonie, 42–74. Cambridge: Cambridge University Press, 1985.

French, Roger K. *Dissection and Vivisection in the European Renaissance.* Aldershot and Brookfield VT: Ashgate, 1999.

French, Roger K. "*De Juvamentis Membrorum* and the Reception of Galenic Physiological Anatomy." *Isis* 70, 1979: 96–109.

Garofalo, Ivan. "Agostino Gadaldino (1515–1575) et le Galien latin". In *Lire les médecins grecs à la Renaissance: aux origines de l'édition médicale: actes du colloque international de Paris, 19–20 septembre 2003*, edited by Véronique Boudon-Millot and Guy Cobolet, 283–322. Paris: Bibliothèque interuniversitaire de médecine, 2004.

Garofalo, Ivan and Armelle Debru. *Galien, Tome VII. Les os pour les débutants, L'anatomie des muscles.* Paris: Les Belles Lettres, 2005.

Garrison, Daniel H. *Andreas Vesalius.* The China Root Epistle. *A New Translation and Critical Edition.* Cambridge: Cambridge University Press, 2015

Gaskell, Philip. *A New Introduction to Bibliography.* Winchester: St Paul's Bibliographies, 1995.

Gaskell, Roger. "The Typography and Layout of Vesalius' De Fabrica as specified in his Letter to Oporinus." *L'Illustrazione, Rivista del libro a stampa illustrato* 3, 2019: 29–53.

Giacomelli, Carlo. "Medicina Patavina. Codici di medicina a Padova, fra Bessarione, Niccolò Leonico Tomeo e Marco Antonio della Torre (?)." *Revue d'histoire des Textes* 16, 2012: 75–113.

Gilias, Guy. "Ornamental letters in the Fabrica 1543 and 1555." In *Art of Vesalius,* edited by Rob Van Hee, 97–112. Antwerp and Apeldoorn: Garant, 2014.

Gilly, Carlos. *Spanien und der Basler Buchdruck bis 1600: Ein Querschnitt durch die spanische Geistesgeschichte aus der Sicht einer europäischen Buchdruckerstadt.* Basle: Helbing und Liechtenhahn, 1985.

Gilly, Carlos. *Die Manuskripte in der Bibliothek des Johannes Oporinus,* Basle: Schwabe, 2001.

Gleason, Maud W. (2009) "Shock and Awe: The Performance Dimension of Galen's Anatomical Demonstrations." In. *Galen and the World of Knowledge,* edited by Christopher Gill, Tim Whitmarsh and John Wilkins, 85–114. Cambridge: Cambridge University Press, 2009.

Gliozzi, Giovanni. "Canano, Giovanni Battista, il Giovane." In *Dizionario biografico degli italiani* (1960-). Rome: Treccani 17, 1974: 714–716.

Goree, John A. "The Woodblocks of Vesalius and the Printings: From the Renaissance to the Modern Era." *Printing History* 16, 2014: 3–41.

Grafton, Anthony. "The Importance of Being Printed." *The Journal of Interdisciplinary History* 11, 1980: 265–86.

Grendler, Paul F. *The Universities in the Italian Renaissance.* London and Baltimore: Johns Hopkins University Press, 2002.

Gross, Dominik and Jan Steinmetzer. *Volcher Coiter (1534–1576) und die Konstituierung ärztlicher Autorität in der Vormoderne.* Aachen: Shaker Verlag, 2006.

Guerra, Francisco. "The Identity of the Artists involved in Vesalius' *Fabrica* 1543." *Medical History* 13, 1969: 37–50.

Gundert, Beate. (2006) "Zu den Quellen der Basler Galen-Ausgabe (1538)." In *Ärzte und ihre Interpreten: Medizinische Fachtexte der Antike als Forschungsgegenstand der Klassischen Philologie,* edited by Carl W. Müller, Carl Brockmann and Carl W. Brunschön, 81–100. Munich and Lepizig: K. G. Saur, 2006.

Gysel, Christoph. "Jan Boekel (1535–1605) en zijn "Anatome" (1585), een plagiaat van Bordings "Physiologia" (1591)." *Acta Belgica Historiae Medicinae* 9(1), 1996: 21–27.

Harthan, John. *The History of the Illustrated Book. The Western Tradition.* London: Thames and Hudson, 1980.

Helm, Jürgen. "Religion and Medicine: Anatomical Education at Wittenberg and Ingoldstadt." In *Confessions and the Sciences in the Sixteenth Century*, edited by Jürgen Helm and Annette Winkelmann, 51–68. Leiden: Brill. 2001.

Herrlinger, Robert. "Die Initialen in der <Fabrica> des Andreas Vesalius." *Acta Medicae Historiae Patavina* 10(1), 1963–1964: 98–116.

Herrlinger, Robert and Feiner, Edith. "Why did Vesalius not Discover the Fallopian Tubes?" *Medical History* 8, 1964: 335–341.

Hieronymus, Frank. *Oberrheinische Buchillustration 2, Basler Buchillustration 1500–1545*. Basle: Universitätsbibliothek, 1984.

Hieronymus, Frank. "Physicians and Publishers in Sixteenth-century Basle." In *The German Book 1450–1750. Studies Presented to David L. Paisey on his Retirement*, edited by John L. Flood and William A. Kelly, 95–110. London: The British Library, 1995.

Hieronymus, Frank. *Theophrast und Galen – Celsus und Paracelsus. Medizin, Naturphilosophie und Kirchenreform im Basler Buchdruck bis zum dreissigjährigen Krieg*. Basle: Universitätsbibliothek, 2005.

Hobson, Anthony. *Renaissance Book Collecting: Jean Grolier and Diego Hurtado de Mendoza, their Books and Bindings*. Cambridge: Cambridge University Press, 1999.

Horowitz, Michael and Jack Collins. "A Census of Copies of the First Editions of Andreas Vesalius' *De humani corporis fabrica* (1543) with a Note on the Recently Discovered Issue." *Journal of the History of Medicine and Allied Sciences* 39, 1984: 198–221.

Houtzager, Hans L. "Editions, Copies and Replicas of *Fabrica* and *Epitome*, published between 1543 and 1725." In *Art of Vesalius*, edited by Rob Van Hee, 190–220. Antwerp and Apeldoorn: Garant, 2014.

Huard, Pierre and Mirko Drazen Grmek. *L'oeuvre de Charles Estienne et l'école anatomique parisienne*. Paris : Au cercle du livre précieux, 1965.

Ieraci Bio, Anna Maria, "Giorgio Valla e la medicina. Disegni anatomici e nomenclatura delle parti del corpo umano nel Mutin.α. W.5.5." *Galenos* 11, 2017: 227–245.

Janssen, Frans A. "The Rise of the Typographical Paragraph." In *Cognition and the Book. Typologies and Formal Organisation of Knowledge in the Printed Book of the Early Modern Period*, edited by Klaus A. Enenkel and Wolfgang Neuber, 9–32. Leiden and Boston: Brill, 2005.

Joffe, Stephen N. *Andreas Vesalius, the Making, the Madman and the Myth*. Bloomington: AuthorHouse, 2014.

Joffe, Stephen N. and Veronica Buchanan. "The Vesalius Epitome of *De Humani Corporis Fabrica* of 1543. A Worldwide Census with New Findings." *Medical Research Archives* 2, 2015: 1–13.

Jones, Peter M. "Thomas Lorkyn's Dissections, 1564/5 and 1566/7." *Transactions of the Cambridge Bibliographical Society* 9(3), 1988: 209–229.

Jurina, Kitti. *Vom Quacksalber zum Doctor Medicinae*. Cologne: Bohlau Verlag, 1985.

Kaap, Henry. "Porta, Giuseppe." In *Allgemeines Künstler Lexikon, Band 96, Pintaldi – Pretro*, edited by Andreas Beyer, Bénédicte Savoy and Wolf Tegethoff, 368–370. Berlin: De Gruyter, 2017.

Kavvadia, Maria. "Sources and Resources of Court Medicine in Mid-Sixteenth Century Rome: Erudition as an Epistemological and Ethical Claim." In *Scientiae in the History of Medicine,* edited by Fabrizio Baldassari and Fabio Zampieri, 171–88. Rome and Bristol, CT: L'Erma di Bretschneider, 2021.

Keele, Kenneth D. and Carlo Pedretti. *The Drawings of Leonardo da Vinci in the Collection of Her Majesty the Queen at Windsor Castle.* New York: Harcourt Jovanovitch, 1978–1980.

Kellett, Cyril E. "Sylvius and the Reform of Anatomy." *Medical History* 5, 1961: 101–116.

Kemp, Martin, "A Drawing for the *Fabrica*: And Some Thoughts Upon the Vesalius' Muscle-men." *Medical History* 14, 1960: 277–288.

Kemp, Martin. *Leonardo: Experience, Experiment and Design.* London: V & A, 2006.

Kemp, Martin. "The Mark of Truth: Looking and Learning in some Anatomical Illustrations from the Renaissance and the Enlightenment." In *Medicine and the Five Senses,* edited by William F. Bynum and Roy S. Porter, 85–121. Cambridge: Cambridge University Press, 1993.

King, Helen. *The One-sex Body on Trial: The Classical and Early Modern Evidence,* Farnham and Burlington, VT: Ashgate, 2013.

Klestinec, Cynthia. "Juan Valverde de (H)Amusco and Print Culture. The Editorial Apparatus in Vernacular Anatomy Texts." In *Zergliederung – Anatomie und Wahrnehmung in der Frühen Neuzeit,* edited by Albert Schirrmeister and Matthias Pozsgai, 78–94. Frankfurt am Main: Vittorio Klostermann, 2005.

Klestinec, Cynthia. "Medical Education in Padua: Students, Faculty and Facilities." In *Centres of Medical Excellence: Medical Travel and Education in Europe, 1500–1789,* edited by Ole P. Grell, Andrew Cunningham, and Jon Arrizabalaga, 193–211. Farnham: Ashgate, 2010.

Klestinec, Cynthia. *Theaters of Anatomy: Students, Teachers and Traditions of Dissections in Renaissance Italy.* Baltimore: Johns Hopkins University Press, 2011.

Kornell, Monique N. *Artists and the Study of Anatomy in Sixteenth-Century Italy.* PhD Diss. University of London, 1993.

Kornell, Monique N. "A Dating for Domenico Bonaveri's *Notomie di Titiano.*" In *Print Quarterly* 33(4), 2016: 379–390.

Kornell, Monique N. *Flesh and Bones: The Art of Anatomy.* Los Angeles: Getty Trust Publications, 2022.

Kornell, Monique N. "Jan Steven van Calcar (c. 1515–c. 1546), Vesalius's Illustrator." In *Andreas Vesalius and the "Fabrica" in the Age of Printing: Art, Anatomy, and Printing in the Italian Renaissance*, edited by Rinaldo Canalis and Massimo Ciavolella, 99–130. Turnhout: Brepols, 2018.

Kornell, Monique N. "Rosso Fiorentino and the Anatomical Text." *The Burlington Magazine* 131(1041), 1989: 842–847.

Kornell, Monique N. "Vesalius' Method of Articulating the Skeleton and a Drawing in the Collection of the Wellcome Library." *Medical History* 44, 2000: 97–110.

Kornell, Monique N. and Daniel Margócsy. "'A Spring of Immortal Colours': Jacques Le Moyne de Morgues (c. 1533–1588) and Picturing Plants in the Sixteenth Century." *Journal of the Warburg and Courtauld Institutes* 86, 2023:109–157.

Kosmin. Jennifer F. "Midwifery Anatomized: Vesalius. Dissection and Reproductive Authority in Early Modern Italy." *The Journal of Medieval and Early Modern Studies* 48, 2018: 79–104.

Kusukawa, Sachiko. "*Ad vivum* Images and the Knowledge of Nature in Early Modern Europe." In *Ad vivum? Visual Materials and the Vocabulary of Life-Likeness in Europe before 1800,* edited by Thomas Balfe, Joanna Woodall and Claus Zittel, 89–121. Leiden and Boston: Brill, 2019.

Kusukawa, Sachiko. *Andreas Vesalius. Anatomy and the World of Books*, London: Reaktion Books, 2024.

Kusukawa, Sachiko. *Picturing the Book of Nature. Image, Text and Argument in Sixteenth-Century Human Anatomy and Medical Biology.* Chicago: Chicago University Press, 2012.

Kwakkelstein, Michael. W. *Study Heads and Anatomical Studies, I. Corpus Rubenianum Ludwig Burchard, Part XX.* London and Turnhout: Harvey Miller Publishers, 2021.

Kyle, Sarah R. *Medicine and Humanism in late Medieval Italy: the Carrara Herbal in Padua.* London: Routledge, 2017.

Lambert, Samuel W. "The Initial Letters of the Anatomical Treatise *De Humani Corporis Fabrica* of Vesalius." In Samuel W. Lambert, Willy Wiegand, and William M. Ivins Jr. *Three Vesalian Essays to Accompany the Icones anatomicae of 1934*, 1–24. New York: The Macmillan Company, 1952.

Landers, Matthew, "Early Modern Dissection as a Physical Model of Organization." In *Anatomy and the Organization of Knowledge, 1500–1800,* edited by Matthew Landers and Biran Muñoz, 9–23. London: Pickering and Chatto, 2012.

Landucci Ruffo, Paola. "Le fonti della medicina nell'enciclopedia di Giorgio Valla." In *Giorgio Valla tra scienza e sapienza*, edited by Vittore Branca, 55–68. Florence: Olschki, 1981.

Lanska, Douglas J. "Evolution of the Myth of the Human Rete Mirabile Traced through Text and Illustrations in Printed Books: The Case of Vesalius and his Plagiarists." *Journal of the scientific History of the Neurosciences* 31, 2022: 221–261.

Laurenza, Domenico. "Art and Anatomy in Renaissance Italy. Images from a Revolution." *Metropolitan Museum of Art Bulletin* 69(3), 2012: 1–48.

Laurenza, Domenico. "Contenuto, canoni visive e techniche nella illustrazioni anatomiche del Rinascimento. Riflessioni in margine ad una tavola di Berengario da Carpi." In *Anatome: sezione, scomposizione, raffigurazione del corpo nell'età moderna*, edited by Giuseppe Olmi and Claudia Pancino, 27–44. Bologna: Bononia University Press, 2012.

Laurenza, Domenico. "In Search of a Phantom; Marcantonio della Torre and Leonardo's late Anatomical Studies." In *Leonardo's Anatomical World. Language, Context and "Disegno"*, edited by Alessandro Nova and Domenico Laurenza, 61–77. Venice: Marsilio, 2011.

Laurenza, Domenico. *La ricerca dell'armonia. Rappresentazioni anatomiche nel Rinascimento.* Florence: L. S. Olschki, 2003.

Lembke, Sven. "Wie der menschliche Leichnam zu einem Buch der Natur ohne Druckfehler wird. Über den epistemologishen Wert anatomischer Sektionen im Zeitalter Vesals." In *Zergliederung – Anatomie und Wahrnehmunge in der Frühen Neuzeit*, edited by Albert Schirrmeister and Matthias Pozsgai, 19–49. Frankfurt am Main: Vittorio Klostermann, 2005.

Leu, Urs B. "Die Bedeutung Basels als Druckort im 16.Jahrhundert." In *Basel als Zentrum des geistigen Austauschs in der frühen Reformation,* edited by Sven Grosse, Christine Christ-von Wedel and Berndt Hamm, 53–78.Tübingen: Mohr Siebeck, 2014.

Lind, Levi R. (translator). *Berengario da Carpi, A Short Introduction to Anatomy.* Chicago: University of Chicago Press, 1959.

Lind, Levi R. (translator). *The* Epitome *of Andreas Vesalius.* New York: MacmIllan Co., 1949.

Lind, Levi R. (translator). *Studies in Pre-Vesalian Anatomy; Biography, Translations, Documents.* Philadelphia: American Philosophical Society, 1975.

Lindberg, Sten G. "Chrestien Wechel and Vesalius: Twelve Unique Medical Broadsides from the Sixteenth Century." *Lychnos* 1953: 50–74.

Lines, David. *The Dynamics of Learning in Early Modern Italy.* Cambridge, MA, and London: Harvard University Press, 2023.

Lo, Melissa. "Cut, Copy and English Anatomy: Thomas Geminus and the Reordering of Vesalius's Canonical Body." In *Andreas Vesalius and the "Fabrica" in the Age of Printing: Art, Anatomy, and Printing in the Italian Renaissance*, edited by Rinaldo Canalis and Massimo Ciavolella, 225–253. Turnhout: Brepols, 2018.

Lonigo, Alberto (ed.) *Nicolò Leoniceno, 1428–1524 Un umanista veneto nella storia della Medicina. Atti del Convegno in Lonigo per il 590 anniversario della nascita.* Contro: Riccardo editore, 2019.

McCall, Taylor. *The Art of Anatomy in Medieval Europe.* London: Reaktion Books, 2023.

MacKinney, Loren C. "The Beginnings of Western Scientific Anatomy: New Evidence and a Revision in Interpretation of Mondeville's Role." *Medical History* 6, 1962: 233–239.

Maclean, Ian. "Vesalius as Authority: from the first Publications to the *Opera Omnia* of 1725." In *Towards the Authority of Vesalius: Studies on Medicine and the Human Body from Antiquity to the Renaissance and Beyond*, edited by Erika Gielen and Michèle Goyens, 23–47. Turnhout: Brepols, 2018.

McNair, Philip. "Poliziano's Horoscope." In *Cultural Aspects of the Italian Renaissance*, edited by Cecil H. Clough, 262–275. Manchester: Manchester University Press, 1976.

Maddison, Francis, Margaret Pelling and Charles Webster (eds). *Essays on the Life and Work of Thomas Linacre c.1460–1524.* Oxford: Clarendon Press, 1977.

Manning, Gideon and Cynthia Klestinec (eds.). *Professors, Physicians and Practices in the History of Medicine. Essays in Honor of Nancy Siraisi.* Cham: Springer, 2017.

Margócsy, Dániel. "From Vesalius through Ivins to Latour: Imitation, Emulation and Exactly Repeatable Pictorial Statements in the *Fabrica*." *Word & Image* 35(3), 2019: 315–333.

Margócsy, Daniel, Mark Somos and Stephen N Joffe. *The* Fabrica *of Andreas Vesalius. A Worldwide Descriptive Census, Ownership and Annotations of the 1543 and 1555 Editions.* Leiden and Boston: Brill, 2018.

Martínez Millán, José. *La Corte de Carlos V, tercera parte, Los Servidores de las Casas Reales*, volume V. Madrid: Sociedad Estatal para la Conmemoración de los Centenarios de Felipe II y Carlos V, 2000.

Maurette, Pablo. "The Organ of Organs: Vesalius and the Wonders of the Human Hand." *The Journal of Medieval and Early Modern Studies* 48, 2018: 79–104.

Meli, Domenico P. "Images & Color: The Strasbourg Printer Johann Schott (1477–1548) and His Circle." *Early Science and Medicine* 27(4), 2022: 527–571.

Menchi, Silvana. "Bizzarri, Pietro." In *Dizionario biografico degli italiani* (1960-). Rome: Treccani 10, 1968: 738–741.

Meyer, Frederick G., Emily E. Trueblood, and John L Heller (eds). *The Great Herbal of Leonhart Fuchs: De historia stirpium commentarii insignes, 1542.* Stanford: Stanford University Press and Cambridge: Cambridge University Press, 1999.

Mondrain, Brigitte. "Éditer et traduire les médecins grecs au XVIe siècle: L'exemple de Janus Cornarius." In *Les voies de la science grecque: Études sur la transmission des textes de l'Antiquité au dix-neuvième siècle*, edited by Danielle Jacquart, 391–417. Paris: De Boccard, 1997.

Montfort, Marie Laure. *Janus Cornarius et la redécouverte d'Hippocrate à la Renaissance.* Turnhout: Brepols, 2018.

Mortimer, Ruth. "The Author's Image: Italian Sixteenth-century Printed Portraits." *Harvard Library Bulletin* 7(2), Summer 1996: 7–87.

Mudry, Albert. "Disputes Surrounding the Discovery of the Stapes in the Mid-16th Century." *Otology & Neurotology* 34(3), 2013: 588–592.

Mugnai Carrara, Daniela. *La biblioteca di Nicolò Leoniceno tra Aristotele e Galeno:cultura e libri di un medico umanista*. Florence : L. S. Olschki, 1991.

Mugnai Carrara, Daniela. "Profilo di Nicolò Leoniceno." *Interpres* 2, 1989: 169–212.

Munk, William. *The Roll of the Royal College of Physicians of London: 1518 to 1700*. London: Royal College of Physicians, 1878.

Munro, Dane. *Memento Mori. A Companion to the Most Beautiful Floor in the World*. Valletta: MJ Publications, 2005.

Muraro, Michelangelo and David Rosand. *Tiziano e la silografia veneziana del Cinquecento*. Venice: Neri Pozza, 1976.

Murphy, Caroline P. *Lavinia Fontana, A Painter and her Patrons in Sixteenth Century Bologna*. New Haven: Yale University Press, 2003.

Murphy, Hannah. A *New Order of Medicine. The Rise of Physicians in Reformation Nuremberg*. Pittsburgh: University of Pittsburgh Press, 2019.

Nelson, Mary-Jo. *An Analysis of Sixteenth century Venetian Title Pages of Medical Books*, Ph.D. Diss., University of Kansas, Laurence, KA, 1976.

Nickel, Tobias Benjamin. *Die Landschaftszeichnungen von Domenico Campagnola (1500–1564)*. Ph.D. Diss., University of Vienna, 2017.

Nicoll, Charles. Leonardo da Vinci. The Flight of the Mind. London: Penguin, 2004.

Nutton, Vivian. "The Anatomy of the Soul in Early Renaissance Medicine." In *The Human Embryo. Aristotle and the Arabic and European Traditions*, edited by Gordon R. Dunstan, 136–157. Exeter: University of Exeter Press, 1990.

Nutton, Vivian. *Ancient Medicine*, ed. 3. London and New York: Routledge, 2024.

Nutton, Vivian. "André Vésale et l'anatomie parisienne." *Cahiers de l'Association Internationale des Études Françaises* 55, 2003: 239–249.

Nutton, Vivian. "An early reader of Vesalius' *Fabrica*." *Vesalius* 3, 1998: 73–74.

Nutton, Vivian, translator. *An Autobibliography by John Caius*. London and New York: Routledge, 2018.

Nutton, Vivian. *Galen: A Thinking Doctor in Imperial Rome*. London and New York: Routledge, 2020.

Nutton, Vivian. "Galen and the Latin *De Voce*: a new Edition and English Translation." In *Nell' officina del filologo. Studi sul testo e i loro lettori per Ivan Garofalo*, edited by Tommaso Raioli and Amneris Roselli. Pisa and Rome: Fabrizio Serra 2022: 141–164.

Nutton, Vivian. *Johann Guinter and Andreas Vesalius, Principles of Anatomy According to the Opinion of Galen*. London and New York: Routledge, 2017.

Nutton, Vivian. *John Caius and the Manuscripts of Galen. Proceedings of the Cambridge Philological Society, Supplementary Volume* 13. Cambridge: The Cambridge Philological Society, 1987.

Nutton, Vivian. "Medicine: the Dresden Galen." In *Bild und Text im Mittelalter,* edited by Karin Krause and Barbara Schellewald, 347–361. Cologne, Weimar and Vienna, Böhlau Verlag, 2011.

Nutton, Vivian. "*Physiologia*: from Antiquity to Jacob Bording." In *Blood, Sweat and Tears – the Changing Concepts of Physiology from Antiquity into Early Modern Europe,* edited by Manfred Horstmanshoff, Helen King and Claus Zittel. Leiden and Boston: Brill, 2012: 27–40.

Nutton, Vivian. "*Prisci dissectionum professores*: Greek Texts and Renaissance Anatomists." In *The Uses of Greek and Latin, Historical Essays*, edited by A. Carlotta Dionisotti, Anthony Grafton and Jill Kraye.111-26. London: The Warburg Institute, 1988.

Nutton, Vivian. "Qui magni Galeni doctrinam in re medica primus revocavit" Matteo Corti und der Galenismus im medizinischen Unterricht der Renaissance." In *Der Humanismus und die oberen Fakultäten, Mitteilungen XIV der Kommission für Humanismusforschung*, edited by Gundolf Keil, Bernd Moeller and Winfried Trusen, 173–184.Weinheim: VDA, 1987.

Nutton, Vivian. *Renaissance Medicine. A Short History of Medicine in the Sixteenth Century.* London and New York: Routledge, 2022.

Nutton, Vivian. "The Rise of Medical Humanism: Ferrara, 1464–1555." *Renaissance Studies* 11, 1997: 2–19.

Nutton, Vivian. "Vesalius in England, 1544–1547." In *In the Shadow of Vesalius*, edited by Rob Van Hee, 93–100. Antwerp and Apeldoorn: Garant, 2020.

Nutton, Vivian. "Vesalius Revised: his Annotations to the 1555 *Fabrica*." In *Medical History* 56, 2012: 415–443.

Nutton, Vivian. "Wittenberg Anatomy." In *Medicine and the Reformation,* edited by Ole P. Grell and Andrew Cunningham, 11–32. London and New York: Routledge, 1993.

Nutton, Vivian with Gerrit Bos. *Galen, On Problematical Movements.* Cambridge: Cambridge University Press, 2011.

Nutton, Vivian and Christine Nutton. "The Archer of Meudon: a Curious Absence of Continuity in the History of Medicine." *Journal of the History of Medicine and Allied Sciences* 58, 2003: 401–427.

Nutton, Vivian and Christine Nutton. "Noël du Fail, Cardano, and the Paris Medical Faculty." *Medical History* 48, 2004: 367–372.

Oldham, John B. "Shrewsbury School Library." *The Library, ser.* 6(14), 1959: 81–99.

Olmi, Giuseppe and Claudia Pancino (eds). *Anatome: sezione, scomposizione, raffigurazione del corpo nell'età moderna*. Bologna: Bononia University Press, 2012.

O'Malley, C. Donald. *Andreas Vesalius of Brussels (1514–1564)*. Berkeley and London: University of California Press, 1964.

Ongaro, Giuseppe. "Bassiano Landi e Andrea Vesalio." *Atti e memorie dell'Accademia patavina di scienze, lettere ed arti* 110(part 2), 1997–1998: 33–54.

Ongaro, Giuseppe. "Il *'De humana historia'* (1542) di Bassiano Landi." *Atti della VI Biennale della Marca per la storia della medicina (Fermo, 29 aprile-2 maggio, 1965)*, 265–278. Fermo: n.p., 1965.

Ongaro, Giuseppe. "La medicina nello Studio di Padova e nel Veneto." In *Storia della cultura veneta. Vol. III.: Dal primo Quattrocento al concilio di Trento*, edited by Gianfranco Folena, 75–134.Vicenza: Neri Pozza Editore, Part 3, 1981.

Palmer, Richard J. "Nicolò Massa, His Family and his Fortune." *Medical History* 25, 1981: 385–410.

Pardo Tomás, José. "Opening Bodies in the New World. Anatomical Practices in Sixteenth-Century New Spain." In *Anatome: sezione, scomposizione, raffigurazione del corpo nell'età moderna*, edited by Giuseppe Olmi and Claudia Pancini (Bologna: Bononia University Press, 2012), 185–202.

Park, Katherine. "The Criminal and the Saintly Body: Autopsy and Dissection in Renaissance Italy." In *Renaissance Quarterly* 47, 1994: 1–33.

Park, Katherine. "The Rediscovery of the Clitoris." In *The Body in Parts: Fantasies of Corporeality in Early Modern Europe*, edited by David Hillman and Carla Mazzio, 171–194. London and New York: Routledge, 1997.

Park, Katherine. *Secrets of Women: Gender, Generation and the Origins of Human Dissection*. London: Zone Books, 2006.

Park, Katherine. "The Criminal and the Saintly Body: Autopsy and Dissection in Renaissance Italy." *Renaissance Quarterly* 47, 1994: 1–33.

Park, Katherine. "The Rediscovery of the Clitoris." In *The Body in Parts: Fantasies of Corporeality in Early Modern Europe*, edited by David Hillman and Carla Mazzio, London and New York: Routledge, 1997: 171–194.

Park, Katherine. *Secrets of Women: Gender, Generation and the Origins of Human Dissection*. London: Zone Books, 2006.

Petherbridge, Deanna. *The Quick and the Dead. Artists and Anatomy,* London: The South Bank Centre. 1997.

Petrucci Nardelli, Francesca. *La lettera e l'immagine: le iniziali "parlanti" nella tipografia italiana (secc. XVI–XVIII),* Florence: Olschki, 1991.

Pettegree, Andrew. *The Book in the Renaissance,* New Haven: Yale University Press, 2010.

Pigeaud, Jackie. "Formes et normes dans le *De Fabrica de Vésale*." In *Le corps à la Renaissance. Actes du XXXe Colloque de Tours,* edited by Jean Céard, Marie-Madeleine Fontaine and Jean-Claude Margolin, 399–422. Paris: Aux Amateurs de Livre, 1990.

Pigozzi, Marinella. "Bologna. Dall'anatomia agli esemplari del corpo." In *Anatome: sezione, scomposizione, raffigurazione del corpo nell'età moderna*, edited by Giuseppe Olmi and Claudia Pancino, 87–116. Bologna: Bononia University Press, 2012.

Pittion, Jean-Paul. *Le livre à la Renaissance. Introduction à la bibliographie historique et matérielle*, Turnhout: Brepols, 2013.

Pozeg, Zoltan I. and Flamm, Eugene S. "Vesalius and the 1543 Epitome of his *"De humani corporis Fabrica librorum"*: A Uniquely Illuminated Copy." *Papers of the Bibliographical Society of America* 103 (2009): 199–220.

Raschieri, Amadeo A. "Giorgio Valla, Editor and Translator of Ancient Scientific Texts." In *Greek Science in the Long Run. Essays on the Greek Scientific Tradition (4th c. BCE–16th c. CE)*, edited by Paula Olmos, 127–149. Newcastle: Cambridge Scholars Publishing, 2012.

Reeve, Michael D. "Celsus." In *Texts and Transmission. A Survey of the Latin Classics*, edited by Leighton D. Reynolds, 46–47. Oxford: Clarendon Press, 1983.

Reske, Christoph. *Die Buchdrucker des 16. Und 17. Jahrhunderts im deutschen Sprachgebiet. Auf der Grundlage des gleichnamigen Werke von Josef Benzing, 2. überarbeitete und erweiterte Auflage*, Wiesbaden: Harrassowitz, 2015.

Rhodes James, Robert. *Henry Wellcome*, London: Hodder and Stoughton, 1994.

Roberts, Kenneth B. and John D. W. Tomlinson. *The Fabric of the Body. European Traditions of Anatomical Illustration.* Oxford: Oxford University Press, 1992.

Robison, Kira. *Healers in the Making: Students, Physicians, and Medical Eduction in Medieval Bologna (1250–1550).* Leiden and Boston: Brill, 2021.

Rocca, Julius. "Anatomy." In *The Cambridge Companion to Galen*, edited by R. James Hankinson. Cambridge: Cambridge University Press, 2008: 242–262.

Rocca, Julius. *Galen on the Brain. Anatomical Knowledge and Physiological Speculation in the Second Century AD.* Leiden: Brill, 2003.

Roth, Moritz. *Andreas Vesalius Bruxellensis.* Berlin: Georg Reimer, 1892.

Roth, Moritz. "Vesaliana". In *Virchows Archiv* 141, 1895: 462–478.

Salas, Luis A. *Cutting Words: Polemical Dimensions of Galen's Anatomical Experiments.* Leiden and Boston: Brill, 2020.

Salviati. Giuseppe *Regola di far perfettamente col compass la voluta et del capitello ionico et d'ogn' altra sorte.* Venice: Francesco Marcolini, 1552.

Saunders, John B. de C. M. and C. Donald O'Malley. *Illustrations from the Works of Andreas Vesalius of Brussels.* Cleveland: World Publishing Co, 1950.

Sawday, Jonathan. *The Body Emblazoned. Dissection and the Body in Renaissance Culture.* London: Routledge, 1995.

Schneeli, Gustav and Paul Heitz. *Initialen von Hans Holbein.* Strasbourg: J. E. Ed. Heitz, 1900.

Schöne, Hermann. *Apollonius von Kitium.* Leipzig: B. G. Teubner, 1896.

Schultz, Bernard. *Art and Anatomy in Renaissance Italy*. Ann Arbor, Mich.: UMI Research Press, 1985.

Schupbach, William. *The Paradox of Rembrandt's 'Anatomy of Dr. Tulp', Medical History, Supplement* 12, London: The Wellcome Institute for the History of Medicine, 1982.

Screech, Michael A. "Histoire des idées et histoire du livre: une optique personnelle et une profession de foi. " In *Le Livre dans l'Europe de la Renaissance. Actes du xxviiie Colloque international d'Etudes humanistes de Tours*, edited by Pierre Aquilon and Henri-Jean Martin, 554–566. Nantes: PROMODIS, 1988.

Semenzato, Camillo. *The Anatomy Theatre. History and Restoration*. Padua: Università degli Studi, 1995.

Shotwell, R. Allen. "Animals, Pictures and Skeletons. Andreas Vesalius's Reinvention of the Public Anatomy Lesson." *Journal of the History of Medicine and Allied Sciences* 71, 2016: 1–18.

Shotwell, R. Allen. "The Illustrated Dissection: Drawings in the Extant Notes from Vesalius' Dissections." In *Towards the Authority of Vesalius: Studies on Medicine and the Human Body from Antiquity to the Renaissance and Beyond*, edited by Erika Gielen and Michèle Goyens, 383–402. Turnhout: Brepols, 2018.

Sigerist, Henry E. "Albanus Torinus and the German Edition of the "*Epitome*" of Vesalius." *Bulletin of the History of Medicine* 14(5), 1943: 652–666.

Simons, Patricia and Kornell, Monique N. "Annibal Caro's After-Dinner Speech (1536) and the Question of Titian as Vesalius's Illustrator." In *Renaissance Quarterly* 41(4), 2008: 1069–1097.

Singer, Charles and Chaim Rabin. *A Prelude to Modern Science: being a Discussion of the History, Sources and Circumstances of the 'Tabulae anatomicae sex' of Vesalius*. Oxford: Oxford University Press, 1946.

Siraisi, Nancy G. *Medicine and the Italian Universities, 1250–1600*. Leiden: Brill, 2001.

Siraisi, Nancy G. *Medieval and Early Modern Medicine. An Introduction to Knowledge and Practice*. Chicago: University of Chicago Press, 1990.

Skaarup, Bjørn O. *Anatomy and Anatomists in Early Modern Spain*. Farnham and Burlington VT: Ashgate, 2015a.

Skaarup, Bjørn O. "The Unexpected Success of a Spanish Anatomy Book: Juan Valverde de Amusco's *Historia de la composicion del cuerpo humano* (Rome, 1556), and its many later Editions." In *Specialist Markets in the Early Modern Book World,* edited by Richrd Kirwan and Sophie Mullins, 123–141. Leiden: Brill, 2015.

Soulier, David. "The Anatomy Theatre in Renaissance Italy: a Space of Justice?" In *The Spaces of Renaissance Anatomy Theater*, edited by Leslie R. Malland, 23–43. Wilmington and Malaga: Vernon Press, 2022.

Soulier, David. *Examen des* Observations anatomiques *de Gabriel Fallope*. Paris: Classiques Garnier, 2023.
Soulier, David. "The Start of Andreas Vesalius' Career at the Imperial Court according to his Letters to Benedetto Varchi". *In the Shadow of Vesalius*. edited by Rob Van Hee, 37–52. Antwerp and Apeldoorn: Garant, 2020.
Soulier, David. *Vésale anatomisé. Réception et diffusion du* De humani corporis fabrica *en Europe (1543–1628)*. PhD Diss. Université Côte d'Azur, 2020.
Steeno, Omer and Maurits Biesbrouck. "Stolen and lost copies of Vesalius's Fabrica." *Acta medico-historica adriatica* 102(2), 2012, 213–236.
Steeno, Omer, Maurits Biesbrouck, and Theodore Goddeeris). "Six Previously Unknown Letters from Andreas Vesalius to Octavius Landus concerning the Accident of the Spanish Prince Don Carlos in 1562." In *In the Shadow of Vesalius*, edited by Rob Van Hee, 53–92. Antwerp and Apeldoorn 2020.
Steinmann, Martin. *Johannes Oporinus, ein Basler Buchdrucker um die Mitte des 16. Jahrhunderts*. Basle and Stuttgart: Helbing & Lichtenhahn, 1967.
Stolberg, Michael. *Gabrielle Falloppia 1522/3–1562. The Life and Work of a Renaissance Anatomist*, London and New York: Routledge, 2023.
Stolberg, Michael. "Learning Anatomy in Late-sixteenth Century Padua." *History of Science* 56, 2018: 381–402.
Stolberg, Michael. "Teaching Anatomy in Post-Vesalian Padua: An Analysis of Student Notes." *The Journal of Medieval and Early Modern Studies* 48(1), 2018: 61–77.
Sugg, Richard. "The Anatomical Web: Literary Dissection from Castiglione to Cromwell." In *Rhetoric and Medicine in Early Modern Europe*, edited by Stephen Pender and Nancy S. Struever. Farnham and Burlington: Ashgate, 2012: 83–103.
Symons, John. *Wellcome Institute for the History of Medicine, a Short History*, London: The Wellcome Institute, 1993.
Tait, Richard. "The Editions of Gabriele Zerbi's *De cautelis medicorum* and their Influence." *Journal of the Warburg and Courtauld Institutes* 83, 2020: 327–336.
Talvacchia, Bette. *Taking Positions. On the Erotic in Renaissance Culture*. Princeton: Princeton University Press, 1999.
Teixeira Alves, Carlos Fernando. "The Teaching of Anatomy the South of Europe: The Case Study of the Universities of Coimbra and Salamanca (XIV–XVI)." In *The Spaces of Renaissance Anatomy Theater*, edited by Leslie R. Malland, 45–68. Wilmington and Malaga, Vernon Press.
Tietz-Conrat, Erika. "Neglected Contemporaary Sources Relating to Michelangelo and Titian." *The Art Bulletin* 25, 1943: 154–159.
Triebs, Michaela. *Die medizinische Fakultät der Universität Helmstedt (1576–1810)*. Wiesbaden: Harrassowitz Verlag, 1995.

Van Glabbeek, Francis and Maurits Biesbrouck. "Giovanni Baptista Canani (1515–1579) and Andreas Vesalius (1514–1564)." In *In the Shadow of Vesalius*, edited by Rob Van Hee, 101–148. Antwerp and Apeldoorn: Garant, 2020.

Van Glabbeek, Francis, Maurits Biesbrouck and Jacqueline Vons. *Myology before Vesalius. Giovanni Battista Canani (1515–1579 n.s.) and his* Musculorum Humani Corporis Picturata Dissectio. Antwerp and Apeldoorn: Garant, 2022.

Van Hee, Rob (ed). *Art of Vesalius.* Antwerp and Apeldoorn: Garant, 2014.

Van Hee, Rob (ed). *In the Shadow of Vesalius.* Antwerp and Apeldoorn: Garant, 2020.

Van Hee, Rob and Maurits Biesbrouck (eds). *Tijdgenoten uit de leefwereld van Andreas Vesalius.* Antwerp and Apeldoorn: Garant, 2018.

Van Leersum, Everett C. and Wilhelm Martin. *Miniaturen das lateinischen Galenos-Handschrift der kgl. Oeffentl. Bibliothek in Dresden Db 92–93 in phototypischer Reproduktion.* Leiden: A. W. Sijthoff, 1910.

Velut, Stéphane. "L'encéphale et ses enveloppes." In *La* Fabrique *de Vesale. La mémoire d'un livre*, edited *by* Jacqueline Vons, 84–104. Paris: Bibliothèque interuniversitaire de Santé, 2016.

Von Hoffmann, Victoria. "Epistemologies of Touch in Early Modern Autopsies." *Renaissance Quarterly* 75, 2022: 542–582.

Vons, Jacqueline. "Un exemplaire du *De humani corporis fabrica* (1543) annoté par Jan Wauters van Vieringen (Ioan. Valterius Viringus), lecteur et traducteur d'Andre Vésale." *Vesalius* 23, 2015: 38–42.

Vons, Jacqueline (ed). *La Fabrique de Vésale. La mémoire d'un livre,* Paris: Bibliothèque interuniversitaire de Santé, 2016.

Vons, Jacqueline. "Les squelettes de Vésale." In *La* Fabrique *de Vesale. La mémoire d'un livre*, edited *by* Jacqueline Vons, 37–64. Paris: Bibliothèque interuniversitaire de Santé, 2016.

Vons, Jacqueline and Velut, Stéphane. *André Vésale. Résumé de ses livres sur la Fabrique du corps humain.* Paris: Les Belles Lettres, 2008.

Waller, Erik. "Eine unbekannte Ausgabe von Vesals *Epitome.*" *Lychnos* 1936: 152–160.

Weber, Giorgio. *Sensata Veritas. L'affiorare dell'anatomia patologica, ancora innominata, in scritti di anatomisti del '500.* Florence: Leo S. Olschki, 2006.

Wiegand, Willy. "Marginal Notes by the Printer of the ICONES." In Samuel W. Lambert, Willy Wiegand, and William M. Ivins Jr. *Three Vesalian Essays to Accompany the Icones anatomicae of 1934*, 27–42. New York: The Macmillan Company, 1952.

Wilson, Leonard G. "Charles Donald O'Malley 1907–1970." *Journal of the History of Medicine and Allied Sciences* 26, 1971: 82–86.

Wilson, Nigel G. *From Byzantium to Italy. Greek Studies in the Italian Renaissance.* London: Duckworth, 1992.

Index

V. Nutton, *Andreas Vesalius and his Fabrica, 1537–1564*, Palgrave Studies in Medieval and Early Modern Medicine,
https://doi.org/10.1007/978-3-031-69565-0

GPSR Compliance

The European Union's (EU) General Product Safety Regulation (GPSR) is a set of rules that requires consumer products to be safe and our obligations to ensure this.

If you have any concerns about our products, you can contact us on ProductSafety@springernature.com

In case Publisher is established outside the EU, the EU authorized representative is:

Springer Nature Customer Service Center GmbH
Europaplatz 3
69115 Heidelberg, Germany

The manufacturer's authorised representative in the EU is Springer Nature Customer Service Centre GmbH, Europaplatz 3, 69115 Heidelberg, Germany. If you have any concerns regarding our products, please contact ProductSafety@springernature.com

Printed and bound by CPI Group (UK) Ltd, Croydon, CR0 4YY
01/10/2026
02225852-0003